Primary Ha

Ella Victoria Dobbs

Alpha Editions

This edition published in 2024

ISBN 9789362098856

Design and Setting By

Alpha Editions
www.alphaedis.com

Email - info@alphaedis.com

Contents

PREFACE

This book is the outgrowth of long experience as a teacher of primary grades, followed by special study of handwork as a factor in elementary education. It is written with three objects in view:

First, to gather into a single volume various methods already in use in the more progressive schools, and for which the best suggestions are scattered through current periodicals:

Second, to organize these methods and present them in a simple form for the use of teachers who have had no special training in handwork processes:

Third, accepting conditions as they exist in the small town school and the one-room country school, as a basis of organization, to offer suggestions which may be easily adapted to the conditions of any school with a view to bringing present practice into closer harmony with the best educational ideals.

No claim is laid to originality, beyond the small details in which one person's interpretation of a large problem will differ from that of another.

The projects here outlined have been tested in the Public Schools of Columbia, Missouri, under conditions which are common to towns of about the same size.

The point of view has been influenced chiefly by the educational philosophy of Prof. John Dewey, especially as expressed in his essay "The Child and the Curriculum." The author wishes here to make grateful acknowledgement to Dr. Dewey, not only for the helpfulness of his writings, but also for the inspiration of his teaching.

Thanks are also due to Dr. Naomi Norsworthy of Teachers College, and to Dean W. W. Charters of Missouri University, for encouragement in planning the book and for criticism of the manuscript. Especial acknowledgment is here made to Prof. R. W. Selvidge of Peabody College for Teachers, formerly of this University, for hearty coöperation and helpful suggestions in working out the problems described in this book, and to the teachers of the Columbia Schools for their most efficient services in testing these problems in their classrooms.

E. V. D.

UNIVERSITY OF MISSOURI,
February, 1914.

CHAPTER I

INTRODUCTION

In setting forth the plan and purpose of this little book the author wishes to lay equal emphasis on its limitations. The outlines and suggestions which follow are designed for the use of grade teachers who have had little or no training in handwork processes but who appreciate the necessity of making worthy use of the child's natural activity and desire to do. The outlines are arranged with reference to schools which are not provided with special equipment and which have scant funds for supplies. The projects require only such materials as empty goods boxes, and odds and ends of cloth and paper, which are easily obtainable in any community. No extra time is required for the work, and it may be successfully carried out by any teacher who is willing to devote a little study to the possibilities of things near at hand.

These outlines do not form a course of study to be followed in regular order nor in set lessons coming at a definite time. They are, rather, a series of suggestions to be used wherever and whenever they will serve a worthy purpose. They are not to be regarded as a *special* subject, having little or no connection with the regular class work, but rather as an illustrative method of teaching the regular subject matter whenever the teaching can be done more effectively by means of concrete illustrations. It is proposed to make greater use of construction as a medium of expression, and place *making* more nearly on a par with talking, writing, and drawing.

Any of the projects outlined may be modified to suit varying conditions, and the emphasis placed according to the needs of a particular class. All the suggestions are given in very simple form, chiefly from the standpoint of the first grade, for the reason that it is easier to add to the details of a simple problem than to simplify one which is complex.

It is not the purpose here to emphasize the training of the hand or the development of technique in handwork processes to the extent commonly expected of a course in manual arts, though considerable dexterity in the use of tools and materials will undoubtedly be developed as the work proceeds. While careless work is never to be tolerated in construction any more than it would be tolerated in writing or drawing, the standard is to be only such a degree of perfection as is possible through a child's unaided efforts. It is proposed to provide him with things to do of such interest to him that he will wish to do his best, and things of such a nature that they will please him best when they are well done, and so stimulate a genuine desire for good work. To this end the suggestions relate to things of immediate value and use to the children themselves, rather than to things commonly comprehended in a list of articles which are useful from the adult point of view.

The work is to be kept on a level with the child's experience and used as a means of broadening his experience and lifting it to a higher level. It must also be kept on the level of his constructive ability in order that he may do things *by himself*, and develop independence through feeling himself master of his tools. Neither patterns nor definite directions are provided for the details of the projects outlined, for the reason that it is desired to make every project a spontaneous expression of the child's own ideas. To this end the outline serves only as a framework, to be filled in as the worker desires. The ready-made pattern implies dictation on the part of the teacher and mechanical imitation and repetition on the part of the pupil,—a process almost fatal to spontaneous effort. While it is possible through a method of dictation to secure results which seem, at first, to be much better than the crude constructions which children are able to work out for themselves, it is only a superficial advantage, and one gained at the expense of the child's growth in power to think and act independently. It is an advantage closely akin to the parrotlike recitation of the pupil who catches a few glib phrases and gives them back without thought, as compared with the recitation of the pupil who thinks and expresses his thoughts in his own childish language.

These outlines are intended not only to emphasize independence in self-expression, but also to foster a social spirit through community effort and develop a sense of responsibility through division of labor. A child's shortcomings will be brought home to him much more vividly if he fails to contribute some essential assigned to him in the construction of a coöperative project, and thereby spoils the pleasure of the whole group, than when his failure affects only his individual effort in a group of duplicate projects.

These outlines are intended also to suggest a method of opening up to the children, in an attractive way, the great field of industry. Their deep interest in playing store leads easily to a study of the source, use, and value of various forms of merchandise and the essential features of various trades and occupations. Problems of this sort are fascinating to children in all the lower grades, are rich in valuable subject matter, and suggest things to do which are both interesting and worth while. Without attempting to exhaust any phase of the subject, they awaken an intelligent interest in the industrial world and tend to stimulate thoughtful observation. They help to give the children correct ideas about industrial processes as far as their knowledge goes, and to create a desire for further knowledge. This general information lays a good foundation for later and more serious study of the industries and the choice of a vocation.

These outlines are offered as a means of bridging the gap between the formal methods and outgrown courses of study still in use and the richer curriculum and more vital methods toward which we are working. Much time must be spent in study and experimentation before a satisfactory reorganization of the curriculum can be worked out. Without waiting until this work shall be wholly completed, it is possible at once to vitalize the most formal course of study through the use of freer methods, which permit and encourage self-directed activity on the part of the pupil. The use of such methods will not only tend to create a deeper interest in school work, but must also help toward the great problem of reorganization, by throwing into stronger relief the strength and weakness of our present common practice.

CHAPTER II

PAPER CUTTING AND POSTER MAKING

Paper and scissors form a fascinating combination to all children, and offer a very direct means of self-expression. In the language of a small boy who attempted to tell how to do it, "You just think about something and then cut out your *think*." The teacher is concerned chiefly with the "think" and the way in which it is expressed. The children are interested in paper cutting chiefly from the pleasure of the activity. Beyond the immediate pleasure in the process, the cuttings are valuable only as they indicate the clearness of the child's ideas and measure his ability to express them. The process is educative only in so far as it helps the small worker to "see with his mind's eye" and to give tangible shape to what he thus sees. It is important, therefore, that the work be done in a way that will emphasize the thinking rather than the finished product.

The first question arising is, To what extent shall a pattern be used? Shall the teacher cut out the object and bid the class follow her example? Shall she display a silhouette or outline drawing of the object she desires the children to cut, or shall they work without any external guide to justify or modify the mental picture? Shall they be given a pattern and be allowed to draw around it?

All of the above methods are used to a greater or less extent. Long experience seems to indicate that the first cutting of any object should be unassisted by any external representation of it whatever, in order that the attention of each child may be focused upon his own mental picture of the object. When he has put forth his best effort from this standpoint, he should compare his cutting with the real object or a good picture of it and be led to see the chief defects in his own production and then allowed to try again.

FIG. 1.—Story of Jack Horner on poster and sand table. Snowflakes in background. First grade. Columbia, Missouri.

For example, after telling the story of Mother Hubbard, the children may be interested in cutting out dogs. No picture or other guide should be used at first, since every child knows something about dogs. The first cuttings are likely to be very poor, partly because the children have not sufficient control over the scissors and largely because their ideas are very vague. In a general comparison of work they will help each other with such criticisms as, "This dog's head is too big." "That dog's legs are too stiff." They are then ready to try again. Only when they have reached the limit of their power to see flaws in their work do they need to compare it with the real dog or its picture. Only after a child has attempted to express his idea and has become conscious in ever so small a degree of the imperfection of his expression will he really be able to see differences between the real object and his representation of it, and thereby clarify his mental picture.

FIG. 2.—Paper cutting. Second grade, Columbia.

The child's imagination is so strong that he is apt to see his productions not as they are but as he means them to be, and he is unable to distinguish between the original and his copy of it. If the picture or silhouette is presented at first, his work becomes to a large extent mere copying rather than self-expression. If the teacher cuts out a dog and displays it as a sample, the class will be apt to see that piece of paper only and not a real dog. If the children are permitted to draw the outline either freehand or around a pattern, still less mental effort is required, and in cutting they see only the bit of line just ahead of the scissors and not the object as a whole.

Such methods (*i.e.* the use of outlines, silhouettes, etc.) will produce better immediate results. It will be easier to distinguish dogs and cats from cows and horses if a pattern is provided, but it will not produce stronger children. Such methods only defeat the chief purpose of the work, which is to stimulate the mental effort required to hold the mental image of the object in the focus of attention during the time required to reproduce it in the material form.

FIG. 3.—Paper cutting. Second grade.

It is also often asked whether the children shall always cut directly and without modification or whether they shall be permitted to trim off the imperfections of their first attempts. While any rule must always be interpreted in the light of immediate circumstances, it is generally best to cut directly, and after noting the defects, cut again. It is then possible to compare the several attempts and see if improvement has been made. Attention should be directed to the most glaring defect only, and an attempt made to correct it. For example, if the dog's head is too large, do not trim down, but cut another dog and try for better proportions. Compare the second attempt with the first, to measure improvement. Even little children can be taught to work in this thoughtful way, looking for the defects in their own work and making definite attempts to correct them. To this end much cutting from an unlimited supply of newspaper or scratch paper will accomplish more than a few exercises in better paper which must be trimmed and worked over for the sake of economy. If little children are allowed to trim off, they are apt, in the pure joy of cutting, to trim too much and lose the idea with which they started—a process which tends to vagueness rather than clearness. To prevent this it is often helpful to preserve both pieces of paper, *i.e.* the cutting and the hole. (See Fig. 4.)

Paper Tearing.—Paper tearing serves many of the same purposes sought in cutting, and has several strong points in its favor. Working directly with the finger tips tends to develop a desirable dexterity of manipulation. The nature of the process prevents the expression of small details and tends to emphasize bold outlines and big general proportions. Working directly with the fingers tends also to prevent a weak dependence upon certain tools and tends to develop power to express an idea by whatever means is at hand.

FIG. 4.—Paper tearing.

Posters.—The term "poster" as here used includes all mounted pictures made by children, such as cuttings, drawings, paintings, and scrap pictures.

A poster may be the work of one child or of a group. A single poster may tell the whole story, or a series of posters may be made to show a sequence of events. A series of posters may be bound together in book form. For poster making single sheets of paper, medium weight and of neutral tone, are needed. The sheets should be of uniform size for individual use so that they could be bound together if desired. For coöperative work and special problems larger sheets will be needed.

SUGGESTED PROBLEMS FOR PAPER WORK

Cutting out Pictures.—This serves well for first effort with scissors. The interest in the picture furnishes a motive, while the outline serves as a guide and allows the attention to be given wholly to the control of the scissors.

Free cutting of single objects—such as animals, fruits, trees, furniture, utensils, etc.—intensifies and clarifies mental pictures and stimulates observation if the child is led to express his own ideas first and then to compare his expression with the original and note his deficiencies. As far as possible choose objects with strong bold outlines for the first attempts. There should be some marked feature, such as Bunny's long ears, which calls for emphasis. To cut a circular piece of paper which might be an apple or a peach, a walnut or a tomato, will not aid much in clarifying a mental picture, while Bunny's long ears, even though crudely cut, will be more deeply impressed on the child's mind.

Illustrations for Stories.—*Single Illustration.*—After a story has been read aloud and the characters and events freely discussed by the class, each child may be encouraged to represent the part which has appealed to him—*i.e.* "cut what he wants to cut." After the cuttings are mounted they will probably form a series which will tell the whole story. When several children illustrate the same feature, it offers opportunity for comparison and judgment as to which ones have told the story most effectively. For example, in the story of the Three Bears, the cuttings may show the three bears in three relative sizes, the three chairs, the three beds, the table, and the three bowls of porridge. (See notes on Criticism.)

FIG. 5.—Free cutting. Third grade. Columbia, Missouri.

Series.—Let each child select the two or three most important events in a story and illustrate these in a single poster or series of posters.

Community Poster.—A long story such as the "Old Woman and the Silver Sixpence" may be illustrated by the class as a whole, each child cutting some one feature. This requires attention to relative proportions so that the parts may be in harmony when assembled. Such posters may be used for wall decoration.

Charts.—Poster making may also include the making of charts containing samples of manufactured articles in various stages of development. For example, a chart on cotton might show raw cotton, cord, thread, cloth of various sorts, lace, paper, and other materials made from cotton. Such a chart might also include pictures of cotton fields, spinning and weaving machinery, and other related features.

Materials.—Too much can scarcely be said in favor of much cutting from an unlimited supply of common wrapping paper, newspaper, or other waste paper, in which the children are entirely unhampered by such injunctions as, "Be careful and get it just right the first time, because you can't have another paper if you waste this piece." The possible danger of cultivating wastefulness is less serious and more easily overcome than the very probable danger of dwarfing and cramping the power of expression. Here, if anywhere, the rule holds good that we learn to do by doing, and abundant practice is essential to success.

Black silhouette or *poster* paper is most effective when mounted, but is too expensive for general use in large classes.

Brown kraft paper and *tailor's pattern* paper serve well for both cuttings and mounts. Both of these papers may be had by the roll at a low cost. The tailor's paper comes in several dull colors, which make good mounts for cuttings from white scratch paper or the fine print of newspaper.

Bogus paper makes an excellent mount and is very inexpensive.

The Pasting Process.—To a large number of teachers the pasting lesson is a time to be dreaded and its results a cause of discouragement. Especially is this true if the class is large and the teacher attempts to have all the class pasting at one time. In many phases of school work it is so much easier to control forty or fifty children if they all act in unison that we are prone to use the method too often and apply it to forms of work much better managed by groups. The process of teaching little folks to paste is greatly simplified by the use of the group method.

FIG. 6.—Free cutting. Fourth grade. Columbia, Missouri.

If the room affords a large table at which a small group may work, the teacher can easily supervise the work of the entire group. If there is no table, the teacher can work with one or two rows at a time or have very small groups come to her desk. The secret of the success of the group method lies in having the rest of the class busy with some occupation sufficiently interesting to prevent impatience while waiting for turns. The command to "fold hands and sit still till your turn comes" is sure to cause trouble, because children are physically unable to obey it.

The most important factor in successful pasting is a liberal supply of waste paper. Each child should be supplied with a number of single sheets of newspaper torn to convenient size, to paste on, each sheet to be discarded as soon as used. This decreases the danger of untidy work. With the cutting laid upon the waste paper, the paste may be spread with brush, thin wood, or thick paper, well out over the edges. As soon as the pasted cutting is lifted the waste paper should be folded over to cover all wet paste and lessen the possibility of accidents. After the cutting is placed upon the mount, a clean piece of waste paper should be laid over it and rubbed until the air is all pressed out and the cutting adheres firmly. The waste paper overlay may be rubbed vigorously without harm, whereas a light touch of sticky fingers directly upon the cutting will leave a soiled spot, if it does not tear the moist paper. If children are carefully taught in small groups to follow this method of pasting, in a fairly short time all but the weakest members of the class will be able to paste neatly without much supervision.

CHAPTER III

BOOKLETS

The making of booklets forms a valuable accompaniment to almost every phase of school work. Even simple exercises, when put into book form, take on a dignity otherwise impossible and seem more worth while. It is impossible to work with much enthusiasm and care on exercises which are destined only for the wastebasket.

The chief value in the making of booklets is lost when they are made for display purposes only. Many difficulties are sure to arise when the teacher, for the sake of her own reputation, sets an arbitrary standard and tries to force every member of the class to meet it. Because of these difficulties many teachers dread and avoid work of this sort, but the trouble lies in our false standards and poor methods rather than in the process itself. When the exhibit idea is uppermost, each page must be examined with great care, done over again and again if need be, until the standard is reached or the patience of both teacher and pupil exhausted. In such a case the work practically ceases to be the child's own. Instead of expressing an idea of his own in his own way, he tries to express the teacher's idea in the teacher's way, and it is not surprising that he fails so often.

The booklet serves its best purpose when it combines both value and need; that is, when it is something which seems worth while to the pupil and when he feels responsible for its success. He should feel something akin to the responsibility one feels in writing an important letter; that is, that it must be right the first time because there is no opportunity to try again and that he cannot afford to do less than his best because what is done will stand.

To "express his own idea in his own way" does not mean that his work is to be undirected or that poor results are to be accepted. It does mean that when an idea and a means of expressing it have been suggested to him, he shall be allowed to do the best he can by himself, and that when he has done his best, it shall be accepted even though imperfect. Under no circumstances should his work be "touched up" by the teacher. If he is not asked to do things which are too hard for him, he will not make many serious errors. If these are wisely pointed out, they will not often be repeated. If his attention is held to one or two important features at a time, each effort will mean some gain.

The making of a booklet in the primary grades should really consist in making a cover to preserve pages already made or to receive pages on certain topics as they are finished. The making of an animal book, for example, might be a continuous process. Whenever a new animal is studied and a cutting or drawing of it made, the new page may be added to the book.

The first books should be picture books only, collections of cuttings, drawings, and mounted pictures. As the children learn to write they may add first the name and then short descriptions of the pictures, the development proceeding by easy stages until their composition work takes the form of the illustrated story.

Books which are a collection of single sheets are, as a rule, most satisfactory in the primary school. The single sheet is much more convenient to use, and there is always an inspiration in beginning with a fresh sheet of paper. It is more difficult to paste cuttings into a book, and if pages are spoiled, the book is spoiled. If separate sheets are used, a poor one may be done over or discarded without affecting the rest.

The making of booklets and posters offers an excellent opportunity for developing artistic appreciation. It is not enough for the teacher to provide only good colors from which the children may choose, and to supervise the spacing of pictures and then flatter herself that because the results are good that the children are developing good taste. Unless they really want the good things, little real gain has been made. Unless they see some reason for the arrangement of a page, other than that the *teacher wants it that way*, little has been accomplished.

The first attempts will show little or no idea of balance or good spacing. The early color combinations are apt to be crude. If the best things they do are praised and their attention is constantly directed to the good points in things about them, they will begin to want those things. They will begin gradually to feel a greater pleasure in a well-balanced page than in one on which big and little pictures are stuck indiscriminately. If they are given all possible freedom in matters of choice, it will be possible to measure their real progress and to know what points need emphasis.

The more accustomed the children are to tasteful surroundings, the easier will be their progress, but whether they come from tasteful homes or the reverse, the process is the same. Real progress will undoubtedly be slow, but it should be upon a sure foundation.

SUGGESTED TOPICS FOR BOOKLETS

Stories.—Series of illustrations either cut or drawn for any of the stories read by the class.

Animal Book.—Cuttings or sketches of animals. The name and short statement of some characteristic may be added by children who are able to write. Trees, flowers, fruits, etc., may be treated in the same way.

A. B. C. Book.—A page for each letter of the alphabet to be filled with pictures and names of objects having the same initial letter.

House Book.—A page for each room, upon which may be mounted pictures of things appropriate to the room. Newspaper advertisements and catalogs furnish abundant material for this problem. The work not only helps the children to classify present knowledge, but offers opportunity for judgment as to arrangement and relative proportions.

How People Live.—A book of pictures of houses in different countries.

Famous Houses.—Pictures of famous buildings and homes of famous people.

What we Wear.—Pictures showing materials from which clothing is made, the methods of production and manufacture.

What we Eat.—Vegetable foods may be grouped as roots, stalks, leaves, seeds, etc. Animal foods may be classified according to the animal from which they are obtained and the part of the animal from which they are cut. Suggestions for cooking may be added.

How we Travel.—Pictures showing vehicles and conveyances of all sorts, classified as ancient and modern, or according to the countries in which they are used, or the motive power, as horses, electricity, steam, etc.

In connection with elementary geography and history, booklets and posters may be made up from pictures cut from discarded papers, catalogs, and magazines, as well as original drawings. A great variety of topics may be profitably illustrated in this way. As, for example, land and water forms, famous mountains, lakes, rivers, etc., products and processes of cultivation and manufacture, famous people, costumes and customs of other times and places, utensils and weapons of earlier times.

Fastenings.—The simplest method of binding single sheets is by means of paper fasteners and eyelets. Though these are not expensive, some schools cannot afford to buy them. Cords may be used in several ways and serve as part of the decoration.

The Simple Tie.—Punch three holes in the margin, at least one half inch from the edge to prevent tearing out. Insert the cord in the middle hole, carry through one end hole, then through the other end hole, then back through the middle and tie. (See Fig. 7.)

Japanese Sewing.—Punch holes at regular intervals, as one inch apart. Sew through first hole twice, making a loop around the back,—repeat the process until a loop has been made for each hole,—carry the cord in and out through the holes back to the starting point, filling in the blank places and making a continuous line, and tie ends together with a small knot. (See Fig. 8.)

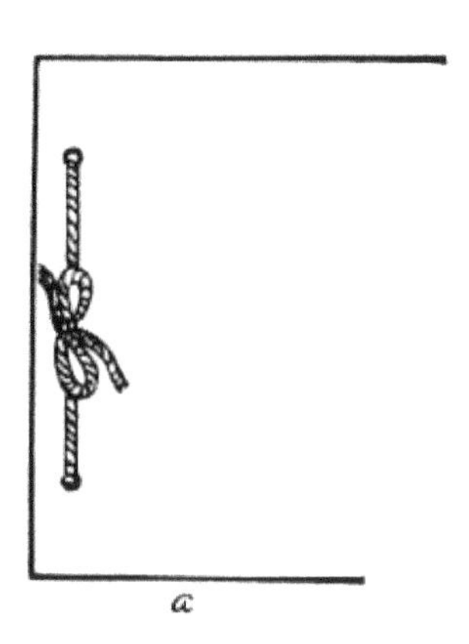

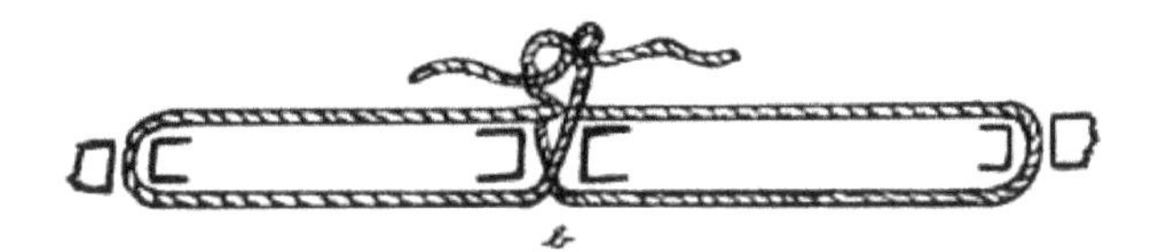

FIG. 7.—Pamphlet sewing.

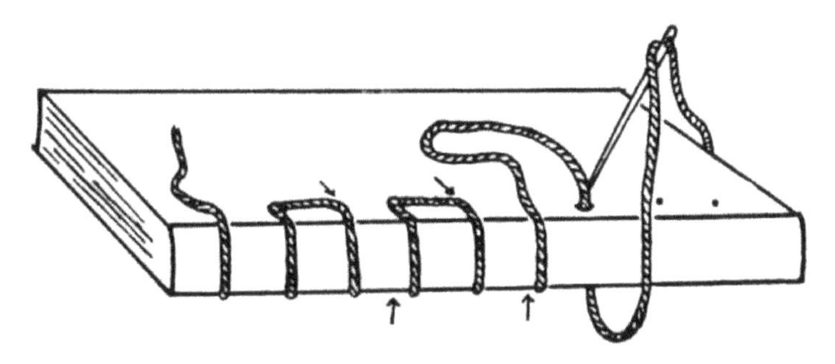

FIG. 8.—Japanese binding.

Decoration.—Only the simplest decoration should be attempted. A plain cover of good color tied with a cord of harmonious color will have elements of beauty without further decoration. A single border line well placed may be used and offers opportunity for developing a nice sense of proportion by studying the results to see which borders are neither too near the edge nor too far from it.

A well-printed, well-placed title is often the most satisfactory decoration. Printing should be introduced early, and the children encouraged to make good plain letters. In order to get the title in good proportion and well placed, it is helpful to cut a piece of paper the desired size and lay it on the cover, moving it about to see where it looks best. Until the children have learned to do fairly neat work it is often helpful to print the title on a separate piece and paste it in place. It is discouraging to spoil an otherwise good cover by a bad letter, and this process lessens that danger.

Before the children learn to print, a simple border or band across the cover may take the place of the title. The border may be drawn in crayons or be free-hand cuttings.

Too much emphasis cannot be laid upon the beauty of simplicity in decoration. Children are inclined to think beauty means fanciness and that beauty increases with the quantity of decoration. It is necessary to begin early to develop a taste for good design.

CHAPTER IV

CRITICISM AND STANDARDS OF WORKMANSHIP

Criticism.—An important feature of all self-directed activity is the ability to judge one's efforts and intelligently measure one's success. This ability is a matter of slow growth and must be cultivated. It is not enough for the teacher to pass judgment upon a piece of work and grade its quality. The worker himself must learn to find his own mistakes and how to correct them. Class criticism offers the best means of developing this power, but must be tactfully conducted.

Little children are brutally frank in expressing their opinions and need to be taught how to be truthful and yet not unkind. They need to be taught what to look for and how to find it, and how to compare one thing with another and discover why one pleases and another displeases. The first essential in the training is emphasis on the good rather than the bad. It is a gospel of "do" rather than of "don't." The earliest efforts of the class may well be confined to comments upon the features they like and, if possible, the reason for the liking. This will forestall any tendency to call undue attention to the poor efforts of weak workers. At first many children will scarcely discriminate between their admiration for a piece of work and their love for the worker and will be apt to praise the work of their special friends. This tendency will gradually disappear through the development of a real basis of appreciation.

The second essential concerns the improvement of the things which are not good. Criticism which merely points out what is bad is of little value. Helpful criticism must point out what is good and why, and what is weak and how to make it stronger. If, for example, the class is considering the success of their efforts to illustrate the story of the Three Bears, they should be encouraged to make such comments as, "John's chairs look too small for his table," "Mary's bowls are all about the same size." The criticism should direct the thought to its possible remedy. It is generally better to pass over defects for which no immediate remedy can be suggested.

Standards of Workmanship.—The standard of excellence by which acceptable work is measured must always vary according to the ability of the class. The best the child can do, alone and unaided, should be the only standard of measurement, and his best efforts should always be accepted, no matter how crude. In no other way can real growth be observed and genuine progress made.

In schools where arbitrary standards are set either by supervisors or by the rivalry of teachers, the tendency to *help* the children by doing part of the work for them for the sake of the *apparent* results, offers the teacher's most serious temptation to selfishness. In a few cases it is helpful for the teacher to add a few strokes to a drawing or adjust some detail in construction, that the child may see the value of certain small changes in the place where they will mean most to him. Such work should not be exhibited as an example of the child's accomplishment, but should be treated as practice work. As a rule the teacher's demonstration should be made on other material and not on that used by the pupil. In no particular are primary schools open to greater criticism than in the too common habit of setting arbitrary standards of excellence and attempting to force all children to reach them. Such standards are usually too high for honest attainment and tempt or force the teacher to use methods which cannot be defended by any sound principle of pedagogy.

Values change with the purpose of the work. A thing is well made when it serves its purpose adequately. Toys must be strong enough to permit handling. Mechanical toys must work. Sewing must be strong as well as neat. In illustrative problems, in which effect is the chief consideration, technique needs little emphasis, and workmanship may be of a temporary character.

Each thing made should establish its own standard in a way to appeal to the child's common sense.

CHAPTER V

THE HOUSE PROBLEM

The making of a playhouse has long been an accepted feature of primary work, but we have not always made it yield all of which it is capable, either in the self-directed activity of the children or in correlated subject matter. It has, in many cases, been only a bit of recreation from the more serious work of the school. In a house prepared by the janitor or older pupils the children have been allowed to arrange and rearrange ready-made furniture contributed from their playthings at home, but little creative work has been attempted. In other cases an elaborate house, carefully planned by the teacher, has been built and furnished by the children, but, because of the detailed planning, the children's part in it became merely a mechanical following of directions. In some cases relative proportions in rooms and furnishings have received scant attention; in others, color harmonies have been all but ignored. These varying methods of carrying out the house-building idea are not without value and may often be justified by local conditions, but their results are meager compared with the possible richness of the problem.

Playing at house building and housekeeping appeals to an interest so universal that children of all times and nations yield to its power. It is therefore necessary to take account of its influence in their development and to dignify it with the approval of the school. We must refine and enrich it by our direction and suggestion without robbing it of its simplicity and charm.

FIG. 9.—Box house, arranged on a shelf.

FIG. 10.—Medieval castle. Built by third grade. Franklin, Indiana.

An example of elaborate work which aroused the interest of pupils and patrons and paved the way for freer work later.

In the suggestions which follow, an attempt is made to utilize this natural activity of children in an occupation which appeals to them as worth while. At the same time it may furnish ample opportunity for the general development and effective teaching of various phases of subject matter which are incident to the occupation, *i.e.* number in connection with measurements, art in the proportions and color combinations, language through discussions and descriptions.

The work is kept on the level of the children's experience by throwing them constantly on their own responsibility in every possible detail, the teacher never dictating the method of procedure and guiding the work with as few suggestions as possible. The work, being on the level of their experience, appeals to the children as very real and worth while. It is, therefore, intensely interesting, and they work without urging.

General Plan.—A house may be constructed from several empty goods boxes, each box forming one room of the house. The boxes or rooms are arranged in convenient order, but are not fastened together. Adjoining rooms are connected by doors carefully cut in both boxes so that the holes match. Windows are also sawed out where needed. The walls are papered, careful attention being given to color schemes, border designs, and relative proportions in spacing. Floors are provided with suitable coverings—woven rugs, mattings, linoleums, tiles, according to the purpose of the room.

Each step is discussed and more or less definitely outlined before the actual making is begun, furnishing opportunity for oral language of a vital sort. Completed parts are examined and criticized, furnishing further opportunity for exercise in oral language while directing attention to strong and weak points in the work.

The materials needed are easily obtainable and inexpensive, consisting chiefly of empty boxes and odds and ends of paper, cloth, and yarn, together with carpenters' scraps.

The tools needed are few, and in some cases may be brought from home by the children for a few days, as needed. The necessary time is found by making the incidental problems serve as subject matter for regular lessons. Making designs for tiling, linoleum, and borders for wall paper, planning relative proportions for doors, windows, and furnishings will supply material for very practical lessons in art. The problems incident to the measurement of doors and windows, tables and chairs, are number work of a vital sort and may be legitimately used as a regular number lesson. Discussions, descriptions, and definite statements of plans all form vital language exercises if rightly used.

HOUSE PLANS IN DETAIL

Materials.—*Empty Store Boxes of Soft Wood.*—Sizes may vary, but where several are grouped for a house, they should be near enough the same height to make a fairly level ceiling. About 10 × 12 × 18 in. is a convenient size.

Paper for Walls.—Scraps of ingrain wall papers may be had from dealers for little or nothing. Cover paper in good colors may be purchased by the sheet. Tailor's paper and brown wrapping paper serve well, and are sold by the roll at a low price.

Pasteboard (strawboard or juteboard) may be used for the roof.

Weaving Materials.—Rugs may be made from carpet rags, rug yarns, rovings, chenille, or jute; towels from crochet cotton; and hammocks from macramé cord or carpet warp.

FIG. 11.—House arranged on a table. Front view. Built by first grade. Columbia, Missouri.

Wood for Furniture.—Bass, white pine, poplar, or other soft wood. Box tops, if of soft wood, may be made to serve nearly all needs. If possible, provide thin wood (about 1/4 in. thick) in various widths, from one inch to six inches, so that only one dimension need be measured. Provide also thick pieces 1 1/2 in. or 2 in. square for beds and chairs; 1/2 in. square for table legs.

Nails of various sizes, chiefly inch brads, are needed.

FIG. 12.—House arranged on a table. Side view. Built by second grade. Columbia, Missouri.

Tools.—The tools actually necessary are few. A class can *get along* with one saw and still do good work, though there will be times when several saws will facilitate progress. Some tools are needed only for a short time and sometimes may be borrowed from the homes. It is more satisfactory to have the school provided with the essential tools whenever possible. The essential tools include:

Brace and auger bit, for boring holes in doors and windows. Needed for a short time only.

Compass saw, for sawing out doors and windows.

Crosscut saw, for sawing off lumber. School should own at least one.

Miter box, for holding lumber and guiding saw. An old one, good enough for children's use, will frequently be contributed by a carpenter. The miter box should be fastened firmly to a low table or box.

FIG. 13.—House arranged on a table. Back view. Built by second grade. Columbia, Missouri.

Hammers, several of medium size.

Try-square, a very valuable tool for setting right angles, provided the teacher and pupils know how to use it.

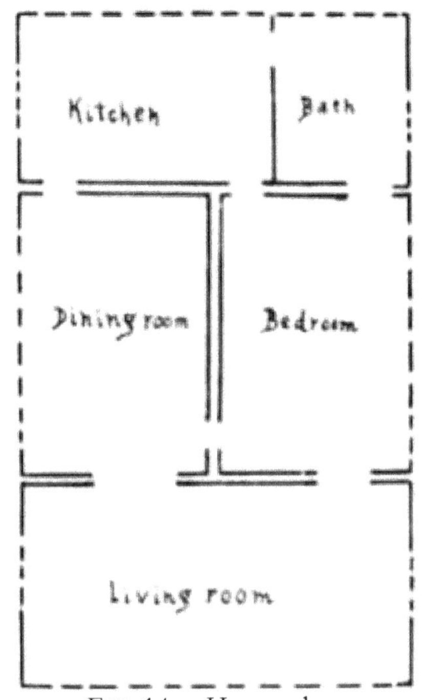

FIG. 14.—House plan.

Arrangement of Rooms.—The sort of house a man can build is governed by his resources and his site. Considering the number of boxes as resources and the table or shelf on which they are to stand as the site, the same big factors which enter into any house-building problem control the size and style of the schoolroom playhouse. What sort of house is desired? What sort of house can be built from the materials at hand? What sort of house can be built in the space at our disposal?

The boxes may be arranged on a shelf with all the open sides toward the class, as in Fig. 9. This economizes space, and all of the rooms are visible at once. A two-story house is easily built on this plan. If economy of space is not necessary, the boxes may be placed on a table with the open sides of the boxes toward the edges of the table, as in Figs. 11, 12, and 13. This permits a more artistic grouping of the rooms. (See Fig. 14.)

The responsibility in grouping the boxes should be thrown as fully as possible upon the children, the teacher merely suggesting where necessary. It should be their house, not the teacher's. The planning should not be hurried but time allowed to discuss the advantages and disadvantages of different plans and reach an agreement. In trying to express individual opinions convincingly their ideas will become clearer—a factor in the development of the children which is much more important than any of the actual details of the house itself. Whether the class decides to have one or two bedrooms in the house is a matter of small consequence. Whether or not they are growing in power to appreciate conditions and make an intelligent decision is a matter of great consequence. Their decisions when made may not always reach the high standard at which the teacher is aiming, but if they have really made a decision, not merely followed the teacher's suggestion, and if their independent selections from time to time show a higher standard of appreciation and greater refinement of taste in ever so small a degree, it is evidence of genuine growth upon a sure foundation.

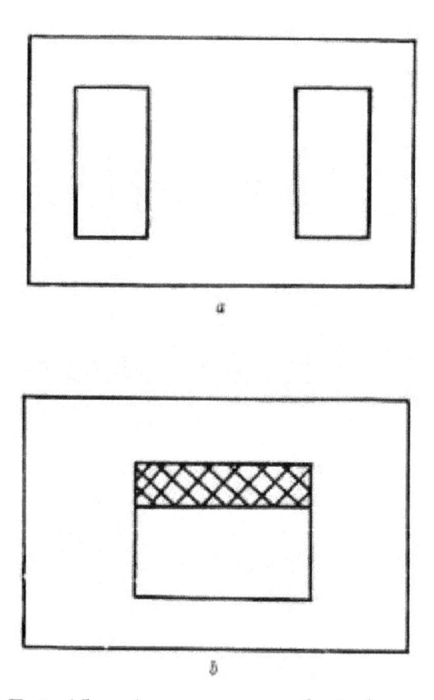

FIG. 15.—Arrangement of windows.

Doors and Windows.—The size and arrangement of doors and windows should be freely discussed. Various possible arrangements may be sketched upon the blackboard by the children. For example, see Fig. 15, *a* and *b*. When a plan is adopted, the doors and windows should be carefully drawn on the *outside* of each box, using the try-square to get right angles.

Bore holes in the corners of the doors and windows and saw out with keyhole or compass saw. In order to avoid mistakes it is well, after sawing out the opening for a door in one box, to place the two boxes together and test the measurements before sawing out the second opening. A mistake of this sort, however, is not fatal, but may prove the most effective way of impressing the workers with the necessity of careful measurement.

Walls.—The decoration of the walls will furnish material for several art lessons. The discussion should turn first to the suitability of different styles for different purposes, such as tiling for kitchen and bathroom walls, light papers for dark rooms, etc. The division of wall space will be the next point to be settled, *i.e.* the height of the tiling or wainscot, the width of a border, or the effect of horizontal and vertical lines in breaking up wall space. These questions may be discussed as far as the immediate circumstances and the development of the class suggest.

The question of color combinations demands special attention. Unless the children come from refined homes their ideas of color will be very crude, and if contributions of material have been asked for, some gaudy impossibilities in flowered paper are apt to be presented. If so, it may require considerable tact on the part of the teacher to secure a satisfactory selection without casting any reflections on the taste of somebody's mother. This difficulty may be avoided to a degree by providing all the materials necessary. It is not enough, however, to cause the children to select good combinations at the teacher's suggestion while in their hearts they are longing for the gaudy thing she has frowned upon. It is better to get an honest expression from them, even though it is very crude, and endeavor to educate their taste to a love for better things, so that each time they choose the choice may be on a higher level of appreciation. Immediate results may not be as beautiful by this plan, and apparent progress may be slow, but only by some such method can a real appreciation be developed which will prevent the return to the crude expression as soon as the teacher's influence is withdrawn.

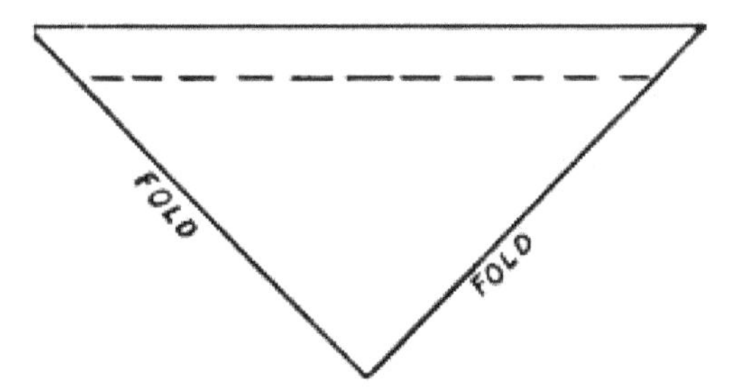

FIG. 16.—Detail of hollow square.

Plain papers generally give the most pleasing effects. Attractive borders may be made by cutting simple units and repeating them at intervals. Almost any motif may be used for the unit. Animals, birds, trees, flowers, ships, etc., serve well. The process of making the border should be a serious lesson in design. A good border is not merely the repetition of a pretty figure. The units must not be too far apart nor too close together. The shape of the figure used must be such that each unit seems to need the next one. Little children will usually take greatest pleasure in working from some nature motif, as flower or animal, but interesting work can be done with simple geometric figures. Take, for example, the hollow square. Fold a square of paper on both diagonals. (See Fig. 16.) Cut on dotted line. Let each child cut several and lay them in order for a border or mount them on a paper of different color. Let the work of the class be put up for general criticism. (See notes on Criticism.) Several points which very small children are able to appreciate will be found to enter into the success or failure of their efforts. The hollow square itself may be cut too wide and look clumsy, or cut too narrow and look frail. In the arrangement they may be too close together and look crowded, or too far apart and look scattered. A sensitiveness to good proportions comes naturally to only a few people, but nearly all are capable of a higher degree of appreciation if their attention is directed to the essential elements which make things good or bad. The beginnings of this appreciation lie in simple things which are easily understood by first-grade children.

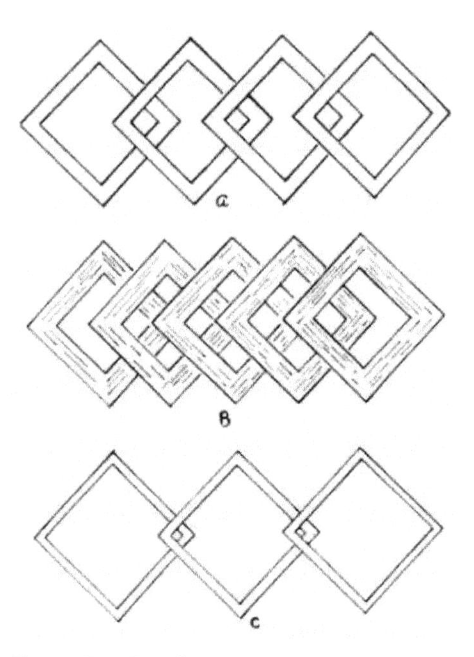

FIG. 17.—Borders using hollow square.

Floors.—Many of the considerations which enter into the selection of wall decorations are of equal importance in choosing floor coverings. What will be suitable to the purpose of each room? Why do we use linoleum in the kitchen and warm rugs in the bedroom? Shall we use small rugs or a carpet? What colors must we have on the floor to harmonize with the colors on the wall? What designs are possible and desirable for the materials we have to use?

Rug Weaving Materials.—The market offers a wide variety of materials prepared especially for school use. Among them the most satisfactory for use with small workers are cotton rovings, loose twisted jute, and cotton chenille. These, especially the first two, are coarse and work up rapidly, and may be had in very desirable colors. Even the cheapest of them, however, will prove an expensive item for the school with limited funds, and ordinary carpet rags may be made to serve every purpose. Often these will be contributed by members of the class. By a careful selection and combination of colors very artistic results can be produced which are in some respects more satisfactory than any obtained from the so-called weaving materials, and have the added advantage of costing practically nothing.

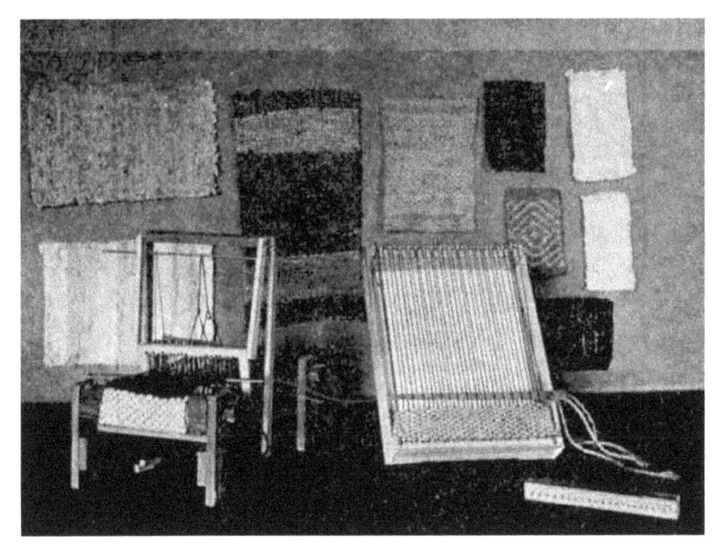

FIG. 18.—Looms and samples of weaving.

Looms.—The market also offers a great variety of looms for school use, many of them quite simple in construction and moderate in price. In schools where bench work is taught, the making of a loom is an excellent problem either for the weavers themselves or for an older class working for them. If the looms are made by the little weavers themselves, only the simplest possible construction should be used, that the work may be completed and the loom put to use before the worker loses sight of the fact that the purpose is to provide carpet for the house. Children lose interest in long-drawn-out processes, and for that reason it is better to provide them with the necessary tools as far as possible while interest in the house building is keen. Later, if considerable enthusiasm has been aroused for weaving, individual looms may be made for home use. For the school with scant funds a very satisfactory loom may be improvised by driving nails one fourth inch apart in the ends of a shallow box of convenient size and stretching the warp threads across the open top.

For very small rugs a cardboard loom will serve. This may be made by cutting notches or punching holes along opposite edges of a piece of cardboard into which the warp may be strung. If a knitting needle is inserted at each side, the cardboard will be stiffened and the edges of the rug kept straight. Weaving needles may be purchased from supply houses. Wooden needles cost 50 cents per dozen. Sack needles serve well for small rugs and may be had at any hardware store for 10 cents per dozen.

Weaves.—For first weaving the plain "over one, under one" on cotton warp with rags or other coarse woof is generally best. Variety may be introduced by weaving a stripe or border of a different tone near each end of the rug. Vertical stripes serve well as another easy method of variation and are produced by using two woof threads of different tones and weaving first with one and then with the other. This weave is very attractive as the body of the rug with a plain border at either end.

As soon as the children have mastered the plain weave and have a fairly clear idea of the possibilities in design through varying the colors in the woof only, they may be initiated into the mysteries of the "gingham weave" and allowed to experiment with the variations in warp as well as in woof. Cotton rovings is an excellent material for weaves of this sort. This weave may also be used with raffia to make matting for the dining-room floor.

FIG. 19.—Box house by second grade. Columbia, Missouri.

Paper mats may also be used as carpets with good effect. Weaving paper strips is often an easier process to little children than weaving with textiles, except where very coarse textile materials are used. For paper mats select paper of suitable color and cut to the size desired for the mat. Fold on the short diameter. Cut slashes from the folded edge, not less than one half inch apart, to within one inch of edge of the paper (See Fig. 20), leaving a margin on all four sides of the mat. For weavers, cut from paper of harmonious tone, strips equal in width to the slashes in the mat.

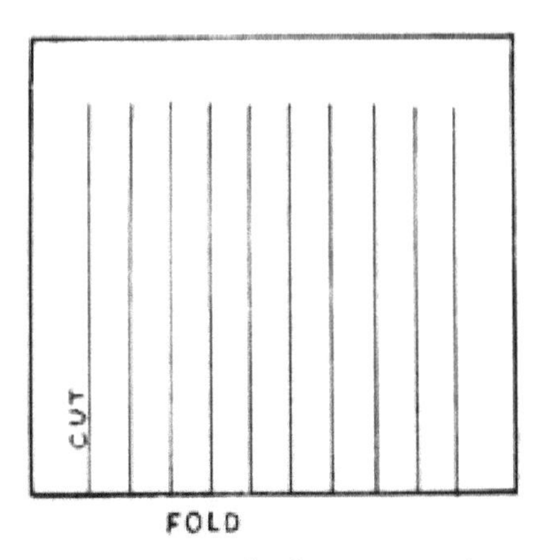

FIG. 20.—Detail of paper weaving.

Variations of the simple over one, under one weave add interest to the work and also give practice in number combinations such as over one, under two, etc. Work of this sort is used in many schools as a method of teaching number, the teacher dictating the combinations while the interest of the children centers in the new pattern which develops under their fingers. While such work has much to be said in its favor, it is open to criticism, especially in the matter of dictation. All the children in any one group will not work with equal speed. Some will undoubtedly "get behind" and others will lose time while waiting for the slow ones. Accidents are liable to happen in individual cases.

Many of these undesirable features may be eliminated while still retaining the valuable part of the work by writing the directions on the board instead of dictating them to the children. It then becomes a lesson in reading as well as in number. Each child is thrown more completely upon his own responsibility and can proceed as rapidly and as steadily as his capacity permits. His rate of progress will often be a fair measure of his ability for independent thought and action, which is the real measure for successful teaching.

As the hardest feature in this method is in keeping the right line and not repeating or omitting any direction, a social spirit may be encouraged by allowing the children to work in groups and take turns in *keeping the place* while the others work. In one first grade where this plan was in vogue the children discovered a book on the teacher's desk which contained numerous designs, many of them much more intricate than she would have attempted to use as classwork. Their instinct for exploration led them to struggle with the directions until they had worked out some designs which would have proved dismal failures had they been attempted as class lessons. In this instance those who belonged to the persevering group were happy in a new-found sense of strength and independence, while the others had accomplished as much as any would have done under the dictation method.

Furniture.—The problem of furniture for the school playhouse has been discussed in numerous publications, and nearly every writer on the subject of primary handwork offers suggestions on this topic. The suggestions include a range in materials and processes from very simple foldings in paper to quite complex processes in reeds and raffia and methodical construction in wood.

Among the various materials and styles in common use, folded paper furniture has the advantage of being quickly made. The process is of sufficient interest to little children to hold their attention, and in order to secure the desired result they must hear the directions intelligently and obey them promptly. These are desirable habits to form. It is quite possible, however, for the work to be done in a very formal, mechanical way, in which the children merely follow directions, often blindly, without any clear purpose and very little thought. Success or failure is due largely to chance; for, if by accident even a good worker "loses out" on a direction, his work is at a standstill until special help is given. He is unable to proceed because he does not know what to do next. There is very little opportunity in such a process for independent thought or action. It is not self-directed activity.

A second objection to paper furniture is its lack of stability. Paper which is pliable enough to fold readily will not hold its own weight long when made into furniture, and very soon becomes wobbly. To overcome this tendency to wobble, heavier papers are often used and new complications arise. Heavy papers do not fold readily without scoring. Scoring demands considerable accuracy of measurement—often to a degree beyond the power of a six-year-old. The stiff papers, being hard pressed, are harder to paste, and neat work is often an impossibility, unless considerable assistance is given.

It is possible to make satisfactory furniture in a great variety of styles from stiff paper, and the processes involve some excellent practice in measurement and design. The processes necessary to obtain these satisfactory results are, however, beyond the ability of children in the lower grades. Even fairly satisfactory results are impossible unless an undue amount of assistance is given by the teacher. In actual practice, where stiff paper is used a few of the best workers in the class are helped to make the few pieces needed in the playhouse and the unhappy failures of the rest of the class are promptly consigned to the wastebasket.

Very pretty furniture may be made from reeds and raffia, but the processes are too difficult to be successfully performed by small children. The reeds do not lend themselves readily to constructions small enough to suit the average playhouse, and the larger pieces are out of proportion to the other features of the house.

The use of wood overcomes the most serious of the objections to be made to other materials, besides being the material most commonly used in "real" furniture. Wooden furniture is stable, and a great variety of processes in construction are possible without introducing complications which prevent independent work on the part of the little people.

The processes necessary to the construction of very simple yet satisfactory wooden furniture may be reduced to measuring one dimension, sawing off, and nailing on. Measuring one dimension is quite within the powers of six-year-olds. *Sawing off* is not difficult if soft lumber is used, and it becomes very simple by the help of the miter box. *Nailing on* is difficult if the nails must be driven into the edges of thin boards, but if thin boards are nailed to thick boards, nails may "go crooked" without serious consequences, and the process becomes quite easy. These processes have the advantage of being particularly fascinating to small boys, in contrast to the girlish character of many forms of primary handwork. (See Figs. 21 and 22.)

FIG. 21.—Furniture from wood blocks.

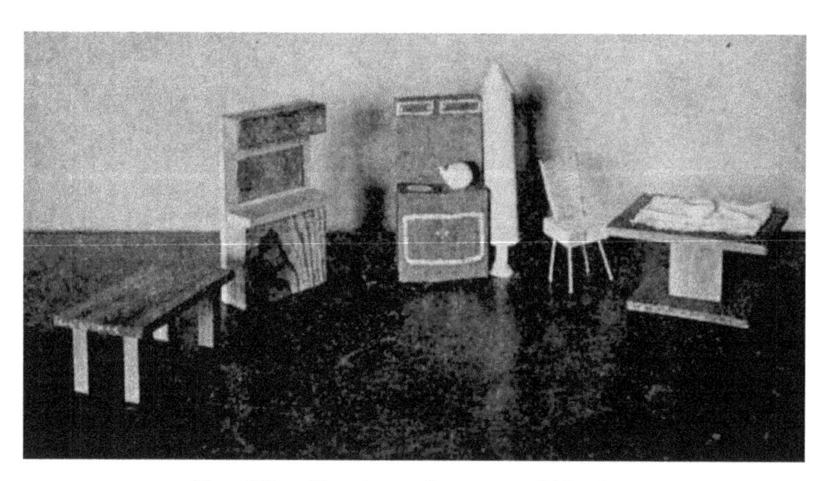

FIG. 22.—Furniture from wood blocks.

Processes.—For the sake of convenience and clearness in these directions it will be assumed that the class is provided with pieces of wood two inches square which will be referred to as 2 × 2. Also with thin wood in a variety of widths from 1 in. to 6 in. Material of other dimensions would serve the purpose equally well, and for many of the parts odd pieces from the scrap box will answer every purpose. The directions are intended only to suggest how to proceed, and it is left to the teacher to adapt them to the material and conditions with which she works.

(1) *To make a chair.*

Use 2 × 2 for seat and thin wood 2 in. wide for back. Children should measure and decide how much to saw off from strip of 2 × 2 in order to make a square block or cube for the seat. They should estimate the length of the back of the chair, then measure and saw off the thin wood needed. Nail the back piece to the cube and finish with a coat of water-color paint or color with crayon. An armchair may be made by the addition of shorter pieces of thin wood to the sides of the chair.

(2) *To make table with pedestal.*

Use 2 × 2 for pedestal. Use thin wood 6 in. wide for top. Use thin wood 4 in. wide for base. Measure and saw off 3 in. of 2 × 2 for pedestal. Measure *enough* of the 6 in. wood to make a square top and *enough* of the 4 in. wood to make a square base. Do not tell the children what they can discover for themselves. They should decide how high the table ought to be and how large to suit the size of the room. Nail the square pieces to the two ends of the pedestal. Finish by same method used for chairs.

(3) *For ordinary table..*

Use thin wood for top. Use ½ × ½ for legs. Measure and saw off pieces needed. Measure places for legs about one inch from corner of top in order to allow an overhang. Children frequently put the legs flush with the edge of the table, which gives a clumsy appearance. Nail through the top with a comparatively long nail.

(4) *To make a double bed.*

Use wood ½ to 1 in. thick for body. Use thin wood of corresponding width for head and foot boards. Class or individual workers should decide on dimensions for different parts and height of body of bed from the floor.

(5) *For single bed.*

Proceed as for double bed, using narrow pieces of wood, or use six or seven inches of 2 × 2 for body of bed and make head and foot boards after the style of chair back.

(6) *Dressing table.*

Decide upon dimensions needed. Use 2 × 2 for body. Use thin wood of equal width for back. Use tinfoil for mirror. Indicate drawers with pencil lines.

(7) *Couch.*

Use piece of 2 × 2 of desired length and make couch cover of appropriate material, or add back and arms of thin wood to piece of 2 × 2 and finish to match other furniture.

(8) *Piano.*

Use wood ¾ or 1 in. thick for body. Nail on piece ½ × ½ for keyboard. Draw keys on paper and paste on keyboard.

(9) *Kitchen stove.*

Use 2 × 4 or any scrap or empty box of appropriate size and shape. Color black with crayon. Add chalk marks or bits of tinfoil to indicate doors and lids. Make hot-water tank of paper. Pieces of reed, wire, or twigs covered with tinfoil make good water pipes. Macaroni sticks and lemonade straws have served this purpose.

FIG. 23.—Home of White Cloud, the Pueblo girl. Second grade. Columbia, Missouri.

Clay Furnishings.—For such articles as the kitchen sink, the bathtub, and other bathroom fittings clay is a satisfactory material. These articles may be modeled by the children, in as good an imitation of the real fittings as they are able to make. Various methods may be used for holding the kitchen sink and the bathroom basin in place, and it is much better for the children to evolve one of their own than to follow the teacher's dictation from the start. If they meet serious difficulties, a suggestion from her may help clear the way. Two long nails driven into the wall will give a satisfactory bracket on which the sink may rest. Two short nails may be driven through the back while the clay is moist and may serve also as a foundation for faucets. The basin, bathtub, and stool may each be built solid to the floor.

The teakettle and other stove furniture may be modeled in clay. Electric light bulbs of clay suspended by cords from the ceiling have a realistic air. Paper shades of appropriate color add to the general effect.

Miscellaneous furnishings.

Bedding.—Paper or cloth may be used for bedding, as circumstances suggest. If interest in *real* things is strong, the making of the sheets and pillow cases offers an opportunity for some practice with the needle. If time is limited, paper may be used.

Curtains.—Curtains also may be made from either paper or regular curtain material. If paper is used, it should be very soft, such as plain Japanese napkins. Scraps of plain net or scrim are most desirable. Some child is apt to contribute a piece of large-patterned lace curtain, but the tactful teacher will avoid using it if possible, and direct the children's thoughts toward a better taste in draperies.

Portières may be made of cloth, of knotted cords, or chenille.

Couch pillows may be made from cloth or may be woven on a small card.

Towels for the bathroom may be woven from crochet cotton.

The fireplace may be made of cardboard marked off and colored to represent brick. A shallow box may be made to serve the purpose. Cut out the opening for the grate and lay real sticks on andirons made from soft wire; or draw a picture of blazing fire and put inside. The fireplace may also be made of clay. Pebbles may be pressed into the clay if a stone fireplace is desired. If clay is used, several small nails should be driven into the wall before the fireplace is built up, to hold the clay in place after it dries.

Bookcases may be made of cardboard, using a box construction, and glued to the wall. Or a block of wood about one inch thick may be used. In either case mark off the shelves and books with pencil lines, and color the backs of the books with crayon.

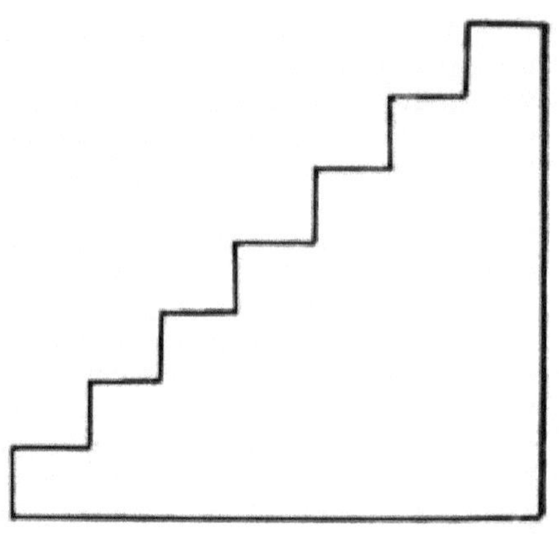

FIG. 24.—Detail of stairway.

The Stairway.—In a two-story house the hardest problem will usually be the stairs. Some good work in number may be done while finding out how many steps will be needed and where the stairway must begin in order to reach the second floor in comfort. Even quite small children can deal with this problem if presented in a simple way. For example, if the box or room is ten inches high, how many steps 1 in. wide and 1 in. high will be needed, and how far out into the room will they come? The children can work out the plan on the blackboard. Measurements may be modified to suit the ability of the class and the needs of the room.

The variety of possible constructions in building the staircase corresponds to the varying ability of classes. A strip of paper may be folded back and forth and made to serve with least mature classes. This paper stair will sag unless it rests on a board or piece of stiff pasteboard. A substantial stairway may be made by sawing two thin boards for supports, as in Fig. 24, and nailing on steps of thin wood or cardboard. There is usually one boy in every first grade who is capable of as difficult a piece of handwork as this. He is apt, also, to be the boy who takes least interest in the general work of the class, and often it is possible to arouse him to special effort through some such problem. The stairway may be made of heavy cardboard with a construction similar to that just described, but this requires pasting instead of nailing and is much more difficult for little children.

The Roof.—The making of the roof is another part of the house building which may often be given into the special care of the two or three over-age pupils who need special problems. The plan which they evolve from their study of the needs of the case will usually be of greater value to them, even though it may not be the best that could be suggested.

The roof may be made of wood as a base, with either wood or cardboard shingles tacked on in proper fashion; or it may be made of cardboard with the shingles merely indicated by lines made with crayon. If the wood base is used, wood gables may be made for sides or ends of the house. To these, long boards may be nailed to form a solid roof. Shingles two inches long by about one inch wide may be cut from cardboard or very thin wood and tacked to the boards. The children should be spurred to study the roofs of houses and find out how the shingles are arranged, and discover for themselves, if possible, the secret of successful shingling.

FIG. 25.—Box house, showing roof. Built by summer class, Teachers Collcgc, New York.

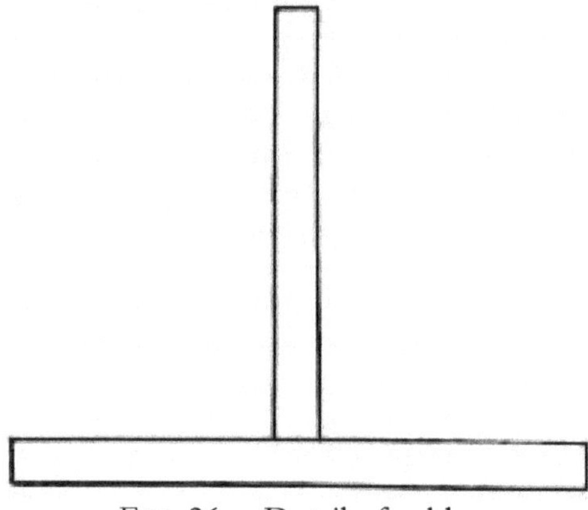
FIG. 26.—Detail of gable.

A cardboard roof is in many ways easier to build. In a house similar to the one shown in Fig. 25 two gables are used, and the roof slopes to front and back. The framework can be very simply made. At the two gable ends place uprights made of two pieces of wood joined in the form of an inverted T. (See Fig. 26.) These should be nailed to the box. A ridgepole may then be nailed to the upper ends of the uprights. If the house is not large, no other framework will be necessary. If the slope of the roof is long enough to allow the cardboard to sag, light strips of wood extending from the ridgepole to the outer edge of the box may be added. If a single piece of cardboard of sufficient size is available, it may be scored[1] and bent at the proper place and laid over the ridgepole, with the edges extending beyond the box to form the eaves. Or, two pieces may be used, one for each slope of the roof, each piece being tacked to the ridgepole. Chimneys may be made from paper and colored to represent bricks or stone.

FIG. 27.—Colonial kitchen. Columbia, Missouri.

The outside of the house may be treated in several ways. It may be sided after the manner of frame houses by tacking on strips of paper or cardboard lapped in the proper fashion. It may be covered with paper marked in horizontal lines to represent siding, in irregular spaces to represent stone, or in regular spaces to represent brick, and finished in the appropriate color. Or, a coat of paint or stain may be applied directly to the box.

VARIATIONS IN HOUSE PROBLEM

A playhouse for its own sake is a justifiable project for primary children and one which may be repeated several times without exhausting its possibilities. Each time it is repeated the emphasis will fall on some new feature, and the children will wish to do more accurate work.

In the lowest grades very simple houses of one or two rooms may be built for story-book friends, such as the "Three Bears" or "Little Red Riding Hood," with only such furniture as the story suggests. In intermediate grades the house may have an historical motive and illustrate home life in primitive times or in foreign countries, such as a colonial kitchen in New England, a pioneer cabin on the Western prairies, a Dutch home, a Japanese home, etc. In upper grades it may become a serious study in house decoration.

As the motive for making the house changes, the character and quality of its furnishings will change. The block furniture described above will give way to more accurate models in either wood or paper. Some excellent suggestions for paper furniture for advanced work may be found in the *Manual Training Magazine.*

As skill in construction increases, a wish for something more realistic than the box construction will arise, and the elements of house framing will be studied with great eagerness.

The House of the Three Bears. (See Fig. 28.)—This house was made early in the year by a class of first-grade children. The walls were papered in plain brown paper. The carpets were woven mats of paper. The chairs, table, and beds were made according to the methods already described in the playhouse outline. The stove and the doll were contributed. The bears were modeled in clay. The children played with the house and its contents throughout the year. The bears were broken and made over many times—a process which not only afforded great pleasure, but also developed considerable skill in modeling.

Another Bears' House.—This house, shown in Frontispiece, was made in the spring, near the end of the school year, by a class of first-grade children all of whom were under seven and many of whom were very immature.

The story of the Three Bears was taken up after Christmas, told and retold, read, and dramatized by the children. Teddy bears were brought to school. Many bears were modeled in clay, each child making the set of three many times.

FIG. 28.—House for the Three Bears. First grade. Columbia, Missouri.

The children laid off spaces on the table for individual Bears' houses and made furniture for these as their fancy prompted. The furniture was made of wood after the general style described above. Later, carpets were woven for these individual playhouses. Each carpet was woven to a given dimension, making it necessary to use the rule. This was their introduction to the rule as a tool for measuring. Every child in a class of forty made one or more pieces of furniture and wove one or more small carpets from rags. Nearly all made some bedding.

FIG. 29.—Cornstalk house. Built by second-grade class. Franklin, Indiana.

Later, four boxes were secured and arranged as a house. The openings for doors were marked off during school time, but were sawed out by a few children who remained during the noon intermission. This is the only part of the work which was not done during regular class time. The papering was done by two or three of the most capable children, while the rest were deeply absorbed in weaving. All made borders. Certain borders were selected for the house, and several children worked together to make enough of the same pattern for one room. Selections were then made from the carpets and furniture already made by the children.

The roof was made chiefly by one boy who "knew a good way to make it." The porches were also individual projects by pupils who had ideas on the subject and were allowed to work them out.

The children became very familiar with every phase of the story and attacked any expression of it with the feeling, "That's easy." They wrote stories, *i.e.* sentences about bears. Each child at the close of the year could write on the blackboard a story of two or more sentences. They made pictures of bears in all sorts of postures with colored crayon and from free-hand cuttings. They modeled the bears in clay over and over again, keeping up a large family in spite of many accidents.

Coöperative Building.—Figures 11, 12, and 13 show three rooms of a four-room house built by the first and second grades working together. The living room and bedroom were furnished by first-grade children. The dining room, kitchen, and bath were furnished by the second grade. Four boxes were used. (See diagram, page 35, Fig. 14.) Each room, except the bath, was a separate box. After a general plan had been agreed upon by the teachers, the boxes were carried to the several rooms and each class worked quite independently. When the rooms were finished, they were assembled on a table in the hall and the roof put on.

FIG. 30.—A flour mill. Built by fourth grade. Columbia, Missouri.

The Flour Mill.—The flour mill, shown in Fig. 30, was built in connection with a study of the general subject of milling by a fourth-grade class. The class visited a flour mill. They were shown the various machines, and the function of each was explained to them. They made hasty sketches of the machines and a rough diagram of their arrangement on the floors. They got the dimensions of the floors and height of the ceiling. An empty box was remodeled to approximate the dimensions of the building. Small representations of the machines were made and placed in the proper relation to each other. No attempt was made to show more than the external proportions in the small representation. The work served its best purpose in keeping the children thinking definitely about what they had seen. The attempt to express their thoughts in tangible form deepened the mental impression, even though the tangible results were crude and lacked many details.

The conveyer being of special interest, two boys worked out a larger model which illustrated the band-bucket process. This is shown in Fig. 30, at the right of the mill. Small cups were made of soft tin and fastened to a leather strap. The strap was fastened around two rods, placed one above the other. The lower rod was turned by a crank fastened on the outside of the box. Two or three brads driven into the lower rod caught into holes in the strap and prevented slipping. The machine successfully hoisted grain from the lower box to one fastened higher up, but not shown in the picture. The model was very crude in its workmanship, but it showed the ability of fourth-grade boys to successfully apply an important principle in mechanics, and it gave opportunity for their ingenuity to express itself. The work was done with such tools and materials as the boys could provide for themselves, and without assistance other than encouraging suggestions from the teacher. This bit of construction accompanied a broad study of the subject of milling, including the source and character of the raw materials, the processes involved, the finished products and their value.

CHAPTER VI

THE VILLAGE STREET

Playing store is a game of universal interest. Making a play store is a fascinating occupation. These are factors which cannot be overlooked in any scheme of education which seeks to make use of the natural activities of children.

The downtown store stands to the children as the source of all good things which are to be bought with pennies. It is usually the first place outside the home with which they become familiar, and its processes are sure to be imitated in their play. In their play they not only repeat the processes of buying and selling, but try to reproduce in miniature what they regard as the essential features of the real store.

If they are allowed to play this fascinating game in school, it may, by the teacher's help, become at once more interesting and more worth while. Curiosity may be aroused through questions concerning what is in the store, where it came from, how it got there, what was done to make it usable, how it is measured, and what it is worth. In seeking answers to these questions, the fields of geography, history, and arithmetic may be explored as extensively as circumstances warrant and a whole curriculum is built up in a natural way. After such study, stores cease to be the *source* of the good things they offer for sale. The various kinds of merchandise take on a new interest when the purchaser knows something of their history, and a new value when he knows something of the labor which has gone into their manufacture.

FIG. 31.—Box house and stores. Grades one, three, and two. Columbia, Missouri.

Being a subject of universal interest, it may be adapted to the conditions of the various grades. It being also impossible to exhaust the possibilities of the subject in any single presentation, it may profitably be repeated with a change of emphasis to suit the development of the class. For example, in the second grade, the study of the street is chiefly a classification of the various commodities which are essential to our daily life, and a few of the main facts of interest concerning their origin. Those a little older are interested in the processes of manufacture and the geography of their sources. In playing store, weights and measures, the changing of money, and the making of bills take on an interest impossible in the old-fashioned method of presenting these phases of arithmetic. Discussions and narratives supply oral language work, and descriptions, letters, and notes provide material for written exercises.

The class may be divided into groups, each group contributing one store to the street, or the attention of the whole class may be centered on one store at a time, as the immediate conditions suggest. If the former method is used, as each store is finished it may be used as subject matter for the entire class, while the important facts concerning it are considered. The first permits a broader scope; the second a more exhaustive study. In either case visits to the real stores studied are important supplements to the work.

FIG. 32.—A village street. Third grade. Columbia, Missouri.

General Directions.—Discuss the stores on a village street. Which are most important? Why? Decide how many stores the class can build, and choose those most necessary to a community.

If self-organized groups[2] are allowed to choose the part they are to work out, both interest and harmony are promoted and leadership stimulated.

Use a box for each store. Each group is usually able to provide its own box. Paper inside of box with clean paper, or put on a coat of fresh paint. Make appropriate shelving and counters of thin wood.

Stock the store with samples of appropriate merchandise as far as possible. Supplement with the best representations the children can make. They should be left to work out the problem for themselves to a large extent, the teacher giving a suggestion only when they show a lack of ideas.

Suggestions for Details of Representation.—*Clay Modeling.*—Clay may be used to model fruits and vegetables, bottles and jugs for the grocery; bread, cake, and pies for the bakery; different cuts of meat for the butcher shop; horses for the blacksmith shop and for delivery wagons. Clay representations may be made very realistic by coloring with crayon.

Canned Goods.—Paper cylinders on which labels are drawn before pasting serve well for canned goods. Cylindrical blocks may be cut from broom sticks or dowel rods and wrapped in appropriately labeled covers.

Cloth.—Rolls of various kinds of cloth should be collected for the dry goods store. Figures may be cut from fashion plates and mounted for the "Ready to Wear" department.

Hats.—Hats may be made for the millinery store from any of the materials commonly used. This is a good way for girls to develop their ingenuity and resourcefulness.

The Store Front.—The front of each store may be made of either wood or cardboard, the spaces for doors and windows being left open that the merchandise may be conveniently handled. Brick or stone fronts, second-story windows, offices, etc., may all be indicated as artistically as the capacity of the class permits by the use of colored crayons. The sign is an important feature and should stimulate an interest in well-made lettering.

Additional Projects.—In addition to representations of retail shops, various industries, such as the carpenter shop, blacksmith shop, flour mill, ice plant, and other familiar industries, may be represented. Coöperative institutions, such as the post office and fire department, should be included in the study.

FIG. 33.—A grocery. Fourth grade.

Excursions.—Wherever possible, the plant should be visited by the class. Before making the visit, the class should discuss what they expect to see, and go prepared to find out definite things. Each child should have at least one question which he is to ask, or one item of information for which he is to be responsible to the class on the return. Often the visit is more worth while to the class after they have tried to make a representation from what they already know and from what they can read on the subject. They are then more conscious of their needs and more alive to the important elements than when they are merely seeing a new thing which is to a great extent foreign to their experience. If they make the visit first, they are apt to feel the need of another when they attempt to work out their representation. If they make a representation first, they are quite sure to be dissatisfied with it and want to make another after they have made the visit. In either case their consciousness of need is a measure of growth.

Correlation.—While the building of a store is in progress the study of the sources and processes of manufacture of the various articles of merchandise will supply valuable subject matter in several fields.

English.—Books containing information on the subject will be read with a definite purpose and more than ordinary interest. Especially if the group method is used, will the members of a group be proud to bring to the class interesting items concerning their particular part of the work. These narratives and descriptions may be made excellent practice in either oral or written English and will be of the sort Dewey characterizes as "having something to say rather than having to say something."

Geography.—This study may also enter as deeply into the field of geography as the development of the class warrants. It will be geography of a vital sort. How these things are brought to us touches the field of transportation, creating an interest in ships and railroad trains, pack mules and express wagons.

History.—The study of the process of manufacture opens up the field of industrial history, and in this, as in the geography, the study is limited only by the capacity of the class.

Number.—In the field of number the possibilities are also unlimited, in studying the weights and measures used for different commodities, the actual prices paid for these things, and the usual quantities purchased.

Playing store will involve the making of bills, the changing of money, and the measuring of merchandise. Different pupils may take turns acting as salesmen or cashier. The common practices of business life should be followed as closely as possible, only in this case each purchaser should make out his own bills. Actual purchase slips may be brought from home and used in number lessons.

An inventory of the stock may be taken and will supply excellent practice in addition and multiplication. After the example of *real* stores, a stock-taking sale at reduced rates may be advertised. The writer answered such an advertisement by a third grade and asked how much could be purchased for one dollar. Pencils were busy at once, and a variety of combinations suggested. One pupil was quickly called to account by his mates for offering only ninety-five cents' worth of merchandise for the dollar. By these and numerous other exercises which will suggest themselves to lively children and wide-awake teachers a vast amount of vital subject matter may be dealt with in a natural way, quite on the level of the child's experience and interest.

FIG. 34.—A grocery. Third grade. Columbia, Missouri.

Art.—The art side also may receive due attention in the general proportioning and arrangement of the stores, in the modeling of certain features from clay, as enumerated above, in the making of labels for boxes and cans, in the writing of signs and advertisements, and in the color combinations. These features are to a great extent incidental to other problems just as the use of good taste is incidental to all the affairs of life and should receive corresponding emphasis.

ILLUSTRATIONS

Figure 32 shows about half the stores built by one third-grade class. Some of the subject matter drawn from the various stories was as follows: in connection with the grocery, a study of the source of various articles of food with oral and written descriptions of processes of manufacture; the common measures used in the grocery, and ordinary amounts purchased.

In connection with the meat market, the names of various kinds of meat, the animals from which they are obtained, and the part of the animal which furnishes certain cuts; as, for example, ham, bacon, chops. The current prices and approximate quantity needed for a meal made practical number work.

The bakery called for an investigation of the processes of bread making and a study of the material used. In all of the processes the teacher had opportunity to stress the necessity for proper sanitation.

In connection with the dry goods store, the distinguishing characteristics of cotton, wool, linen, and silk were emphasized and illustrated by the samples collected for the store and by the clothing worn by the children. Common problems in measuring cloth enlivened the number lessons.

The millinery store disclosed considerable ingenuity in the field of hat manufacture, and a lively business in doll hats was carried on for some time.

In connection with the post office, registered letters, dead letters, money orders, rural free delivery, etc., were discussed, and the advantages of coöperation touched upon.

FIG. 35.—A dry goods store. Third grade.

The other stores of the village street offer further opportunity for becoming better acquainted with the common things which lie close at hand and touch our daily lives.

FIG. 36.—Home in a hot country.

FIG. 37.—Home in a cold country.

CHAPTER VII

SAND TABLES AND WHAT TO DO WITH THEM

A sand table should be considered one of the indispensable furnishings of every schoolroom. Its possibilities are many and varied. It may be used merely as a means of recreation and the children allowed to play in the sand, digging and building as fancy suggests. Or it may be used as the foundation for elaborate representations, carefully planned by the teacher, laboriously worked out by the children, and extravagantly admired by the parents on visitors' day. While both of these uses may serve worthy ends on certain occasions, the most valuable function of the sand table strikes a happy medium between the two, as means of illustrating and emphasizing various features of the daily lessons. In this capacity the laborious efforts of the show problem on the one hand and purposeless play of the other are both avoided. In this capacity the work on the sand table goes along hand in hand with the regular work in geography, history, language, or any subject in which it is possible through an illustration to teach more effectively.

The purpose of this work is not so much to produce fine representations as to help the children to clarify and strengthen their ideas through the effort to express them in concrete form. The value lies in the development which comes to the children while they work. The technique of processes of construction is of secondary importance, though careless work ought never to be permitted. The completed project has little value after it has served its purpose as an illustration and may be quickly destroyed to make way for the next project. For this reason emphasis is laid on the general effect rather than the detail of construction. The work should be done well enough to serve the purpose, but time should not be spent on unnecessary details which do not add to the value as an illustration. In most cases speed is an important element. The project should be completed while the subject it illustrates is under discussion, if it is to be of most service. The first essential is that the work shall be done wholly by the children. The teacher may by skillful questions help them to build up in imagination the project they intend to work out, so that they may work with a definite purpose. She may sometimes suggest improved methods of working out various features when the improvements will add to the value of the illustration, but she should seldom, if ever, plan a project definitely or dictate the method of procedure.

Not least among the possible benefits to be derived from work of this kind is the development of resourcefulness. The necessity for expressing an idea in concrete form with whatever materials are at hand often calls for considerable ingenuity. Ability of this sort will show itself only when the children are expressing their ideas with utmost freedom and feel the responsibility for the success of their work. The more earnestly the children try to express their ideas, the greater will be their development. The teacher should feel that she is hindering the growth of the children and defrauding them of their legitimate opportunity for development when she allows an over-anxiety for tangible and showy results to make her take the responsibility upon herself.

The details of method are best presented through a detailed description of typical illustrations actually worked out in the classroom.

A SAND-TABLE FARM—HOME LIFE IN THE COUNTRY

The study of home life as a general subject will include "our home" and the homes of other people who live under different conditions. To the town child the country will often be somewhat familiar and hold the second place in his interest. In the country school the farm may often be the best place to begin.

Various questions will arise as soon as it is decided to make a sand-table farm, the answers to which will be governed by the habits of the locality. What sort of farm shall we have? Shall we raise stock, fruit, corn, wheat, vegetables, or a little of everything? What shall we need to plant in each case, and in what proportion? How much pasture land shall we need? What buildings? What machinery?

FIG. 38.—A sand-table farm. First grade. Columbia, Missouri.

FIG. 39.—A sand-table farm. Second grade. Columbia, Missouri.

Fences.—As soon as the question of crops and the division of the table into fields is settled, the problem of fencing presents itself. What sort of fence is needed, wire, boards, pickets, rails, or hedge? How far apart shall the posts be set, how tall should they be, and how many will be needed? How many boards? How wide? How long? How many wires?

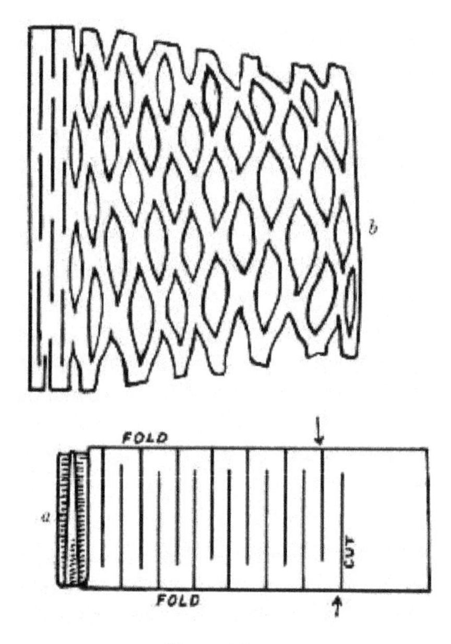

FIG. 40.—
Detail of chicken fence.

The making of the fencing will supply material for one or more number lessons. Various materials may be used.

Twigs may be cut to given lengths and set in concrete (clay) posts.

For wire fence, cut posts from small wooden sticks. Drive small tacks in each post—one for each wire. Use fine spool wire or wire raveled from fly screen. Twist wires once around each tack, or drive the tacks in firmly so that the wire is held by the head of the tack. This is not an easy fence for very little children to make.

To make board fence. Cut posts required length, and decide upon distance between posts. Make boards of thin strips of wood or of pasteboard. Nail boards to posts with tacks or small brads. This is a very easy fence to make and gives some good exercise in measuring.

Rail fences may be made from toothpicks or burnt matches.

Picket fence for the dooryard may be made on wooden foundation with cardboard pickets.

Hedge fence should be made from some fine-leafed plant. Cedar twigs serve well.

Chicken fence may be cut from paper as per illustration. Fold paper several times, lengthwise. Cut across the fold as indicated by arrows. Stretch lengthwise as shown in Fig. 40, *a* and *b*.

Buildings.—The class should decide on the buildings needed. Each building should be assigned to a group of two or three workers. Each group should be held responsible for its contribution and should work out its problem with as little help as possible. If the children are able to plan a barn and make it, even though it is a very crude affair, more has been accomplished than if a very cunning structure had been made after plans, dictated and closely supervised by the teacher.

Wood is the best building material for general use.

Pasteboard serves well, but it is less substantial. It is also harder to cut and paste heavy cardboard than it is to saw and nail thin wood.

Clay may be used for all buildings which are commonly made of concrete.

Stock.—The different kinds of animals needed on the farm and the number of each will furnish profitable subject matter for class discussion. The animals may be modeled from clay. While the animals will of necessity be very large in proportion to the acreage of the farm, attention should be directed to the relative proportions between horses and hogs, cattle and sheep. Differences of this sort do not trouble little people, as their work is sure to show. The point should be stressed only sufficiently to help them to see a little more clearly and express their ideas a little more adequately each time they try. The accuracy of the result is important only as an index that the children are steadily developing in power to see and do, and gaining self-reliance.

The Modeling Process.—The best method seems to be simply to *begin*, and, for example, model as good a horse as possible; then discuss the results, note a few serious defects, and try again, endeavoring to correct them. Encourage rapid work which gives the general proportions of the animal in the rough. Beginners are apt to waste time in a purposeless smoothing of the clay, in mere tactual enjoyment. Discourage the tendency to finish the details of a horse's head, for example, before the body has been modeled. Repeat the process as often as time and the interest of the children warrant, but be satisfied if the children are doing the best they can, even though the results are crude and not so good as some other class has produced. The children should always feel that the work is their own. For this reason the teacher's help in clay modeling should be through demonstration rather than by finishing touches to the child's work. Imitation is a strong instinct in little children, and watching the teacher model a better horse than he can make will help a child to improve his own. One thing to be especially avoided is the attempt to bring every class to a uniform degree of excellence according to adult standards. Such an ideal encourages the giving of help in a way which hinders real development though it may produce immediate results.

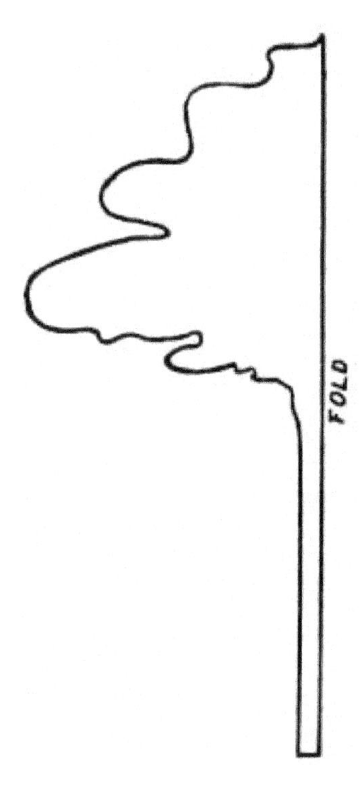

FIG. 41.—
Detail of paper tree.

Trees.—This topic will call out a discussion of the uses of trees; which trees are shade trees, which are cultivated for their fruit, the distinguishing characteristics of the different varieties, and the ones best suited to this particular farm.

Twigs from the real tree should be used wherever possible. In other cases the trees may be cut from paper. If a good green paper is not at hand, use drawing paper and color with crayons. A realistic effect is gained by cutting the tree from folded paper. (See Fig. 41.) Cut three pieces for each tree and paste together at the fold, then open out. Make the trunk long enough to be driven an inch or more into the sand.

The making of the trees will furnish material for both art and nature study lessons. As far as circumstances permit the real trees should be studied, giving the children first-hand experience whether it be much or little. They should test the trees they cut by comparing them with real trees of the same variety. If this is impossible, the best pictures available should be used. (See notes on paper cutting.)

Crops.—When the various parts of the farm are about ready, the fields may be sown. The sand should be made very wet before the seed is put in and sprinkled frequently (twice a day), as the top dries off very quickly. After the seeds have germinated little sprinkling need be done, as the roots will find enough moisture in the wet sand underneath, and it is desirable to retard rather than hasten growth. If carefully managed, a table can be kept green for several weeks.

For corn, check holes well into the sand and drop one grain into each hole. See that rows are straight and holes evenly spaced.

Sow wheat, oats, barley, etc., *very thickly*, cover lightly with dry sand, and sprinkle.

Timothy serves well for meadow and lawn, as it puts up a fine blade. Blue grass sends up a fine blade, but is very slow in germination. Clover does not make a velvety lawn, but a little in the pasture will make an interesting contrast.

Vegetables may be planted in the garden. They will not develop to any great extent, but will serve to emphasize different habits in germination; as, for example, the contrast between beans and corn.

Correlation.—The opportunity for nature study afforded by the farm problem will prove one of its most interesting and valuable features as the progress in plant growth is noted from day to day. The farm problem combines well with both language and art work in supplying vital material for both. In addition to the interesting discussions which naturally arise concerning the building and planting, a diary may be kept by each child.

Keeping a Diary.—The date of planting may be noted and the date when each variety of seed first appears above ground. With the larger seeds, as corn and beans, a seed may be dug up each day and examined, so that the children may appreciate what is going on below ground. Drawings may be made of the seeds, showing the changes in appearance from day to day. After the seed leaves appear the daily growth may be measured and noted in the diary. After a few days seeds may be dug up again that the roots may be examined. At various stages of growth different varieties of seeds may be dug up, laid upon a paper, and sketched by the children. The facts they note may be stated in simple, well-formed sentences, either oral or written or both.

FIG. 42.—An Eskimo village and The Overall Boys' Farm. First grade. Columbia, Missouri.

Art.—The sketching will serve well as the day's art lesson, though its chief value is in helping the children to see clearly. Their efforts will be crude but the teacher should constantly keep in mind that the chief aim is not to obtain fine sketches. Its purpose is to help the children to a better appreciation of the plant through the effort put forth in making the sketch. The technique of the drawing should be emphasized only so far as it will help them express better what they see, and not to the point where they attempt to copy the teacher's strokes. The teacher should be satisfied if every child is doing his best and making steady progress, even though that best may be crude and not up to the standard reached by the teacher who struggles for fine results.

FIG. 43.—An apple orchard. First grade. Columbia, Missouri.

English.—For children who are able to write the diary offers a natural means of gaining experience in the use of common forms of punctuation; as, for example, the writing of dates and the use of a comma in a series, as well as the punctuation of simple statements, in such entries as the following:

April 15, 1912.

We planted the seeds on our farm to-day.

We planted corn, wheat, oats, and beans.

In all work of this sort it is difficult to overestimate the advantage of separate sheets of paper over a notebook with sewed leaves, in the hands of the children. With the fresh sheet always comes an inspiration, no matter what failures have gone before. Poor pages can be done over when necessary, but do not haunt the workers with their discouraging suggestions, as in the use of a notebook. The leaves may be gathered together into a binding of some sort. Even covers of plain brown wrapping paper can be made artistic with a simple border line well placed or a design cut from a paper of a different tone. Written work which culminates in an attractive booklet, however simple, seems more worth while than exercises written into a commonplace notebook or on scratch paper which goes to the wastebasket soon after the mistakes have been commented on.

Number.—The farm problem also supplies abundant opportunity for gaining experience with number. In addition to the actual measurement of the materials used for fences and buildings, the scope may be widened, where conditions warrant, to include estimates and calculations of the amount of the material used.

For example, how many inches or feet of wire will be needed to make a three-wire fence of given length? How large a piece of cardboard will be needed to cut boards one fourth or one half inch wide for a four-board fence fifteen inches long?

These estimates may be translated, *as far as the children are able to appreciate the connection*, into quantities and values of the same material in real problems connected with real farms. It is important, however, to be careful not to carry work of this sort so far beyond the experience of the children that it becomes wholly foreign and abstract to them. We are too apt to forget that it is *experience* and not *objects*, which is the vital factor in concreteness.

FIG. 44.—Robinson Crusoe. Third grade. Columbia, Missouri.

In connection with the nature study a variety of number exercises grow out of the questions which the situation prompts. As, for example, in connection with the corn crop: How many seeds were planted? In how many rows? How many seeds in a row? How many came up? How many failed to germinate? How many more came up than failed? If each good seed should produce two ears of corn, how many would we have? What would they be worth at a given price? etc.

FIG. 45.—Pueblo Indian village. Second grade. Columbia, Missouri.

In an ungraded school, while the younger children might confine their efforts to counting as above, the older children might answer the same questions in terms of percentage and in the probable quantities on a real farm. The stock farm may be treated in the same way. How many cows? How much milk will they give? What will it be worth? How much butter would it make? What will it cost to keep the cows? What is the farmer's profit? These and many other questions will suggest themselves to both teacher and pupils, once the subject is opened up. They will be *practical questions in so far as they touch the experience of the children* in such a way as to appeal to them as real questions. Each individual teacher must decide how far and into what field it is worth while to lead any particular class.

The Sand Table.—The various types of sand tables range all the way from the hardwood, zinc-lined article, provided with a drainpipe, down to the homemade structure evolved from a goods box.

The quality of the table does not greatly affect the quality of the work to be done on it, but there are several points which affect the convenience of the workers. The height of the table should allow the children to work comfortably when standing beside it. A long, narrow table is seldom as satisfactory as one more nearly square, but it should never be too wide for the children to reach the center easily. Any table with tight joints in the top and four- or five-inch boards fitted tightly around the edge will serve the purpose. The inside of the box should be painted to prevent warping and leaking. An "ocean blue" is a good color, as it makes a good background for islands.

If no table is available, a goods box may be turned on its side, the top covered with oilcloth, and a frame, made from the cover of the box, fitted around the edge. The inside of the box may be used as a closet in which to store tools and materials, and a neat appearance given to the whole by a curtain of denim or other plain, heavy material.

ILLUSTRATIVE PROBLEMS

One of the most valuable uses of the sand table is in making illustrations for stories, historical events, and similar topics in which the relations between people and places is important. No definite rules can be laid down for working out such illustrations. The conditions under which they are made, the time to be devoted to the work, the importance of the subject, all affect both the nature and the quality of the work. Any material which lends itself to the purpose should be called into service.

The method of procedure is best set forth by describing several problems as actually worked out by children.

FIG. 46.—A home in Switzerland. Second grade. Columbia, Missouri.

(1) **Story of Columbus**—*First Grade.*

Materials Used.—Paper for cutting and folding, twigs for forests, acorns for tents, large piece of glass for ocean.

Details of Illustration.—The piece of glass was imbedded in sand in the middle of the table; one end of the table represented Spain, the other, America. The representation of Spain included:

"Castles in Spain" being large houses with many windows in which the king and queen lived. They were cut from paper.

Many people, cut from paper, including kings and queens and the friends of Mr. Columbus who came to tell him "good-by." The kings and queens were distinguished by royal purple robes and golden crowns and necklaces, produced by the use of colored crayon.

The three ships made from folded paper. In one of them sat Mr. Columbus.

Fishes, of paper, inhabited the hollow space underneath the glass.

The forest primeval was shown on the American side by green twigs of trees set very close together. On pulling apart the leaves and peering into the depths of this forest, one found it inhabited by bears and other wild beasts, also cut from paper.

The Indians lived in a village of acorn tents set up in a little clearing on the shore.

Flags.—The Spanish region was identified by a Spanish flag, while the stars and stripes waved above the Indian village.

Values.—The project being on the level of the children's experience, they worked freely and with intense interest. The characters in the story were all very real to them. They literally swarmed about the table whenever opportunity was given, moving the figures about as they told the story over and over again. Mr. Columbus sailed across the sea many times. Many boats were made and named for one of the three, according to the preference of the maker. They peeped into the forest and shuddered in delightful fear "lest a bear get me." They made and remade the scene as new ideas suggested themselves during several days of Columbus week.

FIG. 47.—Two little knights of Kentucky. Fourth grade. Columbia, Missouri.

FIG. 48.—How Cedric became a knight. Fourth grade. Columbia, Missouri.

FIG. 49.—A sugar camp. Built late in the spring by a third-grade class. They enjoyed the green grass, though it suggests an overlate season.

Several discrepancies existed which are mentioned here because they troubled some overconscientious visitors. The stars and stripes did not come into existence until centuries after Columbus died and therefore never waved over the Indian village which he found. But chronology does not trouble the first grader very much, while "my country" and "my flag" are ideas which are developing together. And when he is singing, "Columbus sailed across the sea, To find a land for you and me," the red, white, and blue forms the most fitting symbol in his representation of that land. The wild animals which infested the sand-table forest are not all mentioned in the histories as found on San Salvador, but they did exist in the child's idea of the wild country which the white men found on this side of the Atlantic. The children having truthfully expressed their ideas, the teacher had a basis from which to develop, correct, and emphasize such points as were of real importance, while the unimportant features would fade out for lack of emphasis.

FIG. 50.—A western cattle ranch.

On the occasion of the supervisor's visit the members of the class vied with each other in telling the story and explaining the significance of the various illustrations. The supervisor expressed a wish to own some of the cuttings, whereupon, at a hint from the teacher, the class which had gathered about the sand table scampered joyfully (but quietly) back to their seats. Scissors and paper were quickly distributed, and in about five minutes an empty shoe box was required to hold the collection of "Mr. Columbuses," kings and queens in royal purple, gold crowns, and necklaces, ships, fishes, etc., that were showered upon the guest. Needless to say many scraps of paper had fallen to the floor. The teacher remarked that it was time for the brownies to come. Down went all the heads for a sleepy time. The teacher slipped about, tapping here and there a child, who quickly began gathering up the scraps as joyously as he had helped to make them.

The supervisor bade them good-by, with a wish that all children might begin their school life under such happy and wholesome influences.

(2) **Story of Jack Horner**[3]—*First Grade.*—As the story was read the different characters were subjects for free paper-cutting exercises. An abundance of paper (scratch paper and newspaper) was supplied, and each child allowed to cut each figure many times, very quickly.

The story was also dramatized and acted out over and over again. Figure 1 shows the result of an hour's work in assembling the various characters and telling the whole story on the sand table and in a poster. The different figures to be cut were assigned to or chosen by the different children, the teacher taking care that no characters were omitted. Having cut figures of the various characters as they were met in the story, all were eager to reproduce the part called for, and in a few minutes more than enough cuttings were made to supply both sand table and poster with ample material. Two groups of children, one for the poster and one for the sand table, were assigned the work of placing the figures. The teacher superintended both projects, giving a few suggestions as needed, but throwing the responsibility upon the children as much as possible.

This problem was worked out by the same class which made the Columbus illustration just described. The Jack Horner story was illustrated in the spring, after much work of this sort had been done. The quality of the cuttings showed an interesting improvement over the cuttings made for the Columbus story, which came during the third week of the school year.

(3) **Story of Three Little Pigs.**—This is a long story, and three weeks were occupied in reading it and dramatizing it. During this time there were frequent discussions about how it was to be worked out on the sand table. Contributions in great variety were brought in: straw for the straw house, twigs for the house of sticks, bags of brick dust to make a roadway different from the sand, rose hips to be tied to a small branch to represent the apple tree, and various other articles.

FIG. 51.—The story of Three Little Pigs. First grade. Columbia, Missouri.

FIG. 52.—A Japanese tea garden. Third grade. Columbia, Missouri.

FIG. 53.—A coal mine. Fourth grade. Columbia, Missouri.

The houses were built as suggested by the pictures in the reader. The pig and wolf were modeled in clay, each being shown in the several different positions described in the story. Over and over a little clay pig rolled down the hill in a paper churn and frightened a clay wolf. One group, not having wherewithal to build a brick house, used a wooden one made by another group. Another class made the brick house out of blocks, and built in a fireplace with its kettle ready to hold the hot water whenever the wolf should start for the chimney. (See Fig. 51.)

(4) **Japanese Tea Garden.**—A third-grade class used the sand table to illustrate what they had gleaned from reading several stories and descriptions of life in Japan, in connection with elementary geography. The sand-table representation included a tiny bridge across a small stream of "real" water. The "real river" was secured by ingenious use of a leaking tin can which was hidden behind a clump of trees (twigs). A thin layer of cement in the bed of the river kept the water from sinking into the sand. A shallow pan imbedded in the sand formed a lake into which the river poured its waters. (See Fig. 52.)

(5) **A Coal Mine.**—The sand table shown in Fig. 53 was worked out by a fourth-grade class in connection with the geography of the western states. Descriptions and pictures were studied with great earnestness to find out how to fix it, and the children made it as they thought it ought to be. The actual making occupied very little time, the various parts being contributed by different pupils.

Problems of this sort develop leadership. There is usually one whose ideas take definite shape promptly and whose suggestions are willingly followed by his group. If there is one pupil in the class whose ability to lead is so strong that the others are overshadowed, it is sometimes well to let the work be done by small groups who use the table turn about. This plan stimulates a wholesome rivalry and discourages dawdling.

(6) **Stories.**—Illustrations for two stories are shown on page 94. In the first (Fig. 47) part of the class made a representation on the sand table while the rest prepared a poster from paper cuttings. In the second (Fig. 48) empty

shoe boxes were used in making the castle. Very little time was spent on either project.

FIG. 54.—A chariot race. Second grade. Pasadena, California.

CHAPTER VIII

ANIMALS AND TOYS

The circus and the zoölogical garden are always centers of interest to little children and may be used to great advantage to furnish the point of departure in the study of animal life. Making the animals in some form crystallizes the interest in the animals represented, and awakens interest in their habits and home.

The handwork may be used as an illustrative factor connected with geography and nature study, or the making of the circus may be the starting point, and incidentally furnish subject matter in several fields. For example, geography and nature study grow out of the search for facts concerning the animals themselves, *i.e.* size, color, food, home, value, etc. The desire for such information gives purpose to reading. Oral and written descriptions supply subject matter for practice in English. Reducing the actual proportions of animals to a definite scale and problems relating to their commercial value make practical use of the knowledge of number. Art enters into the making of free-hand sketches, cuttings, and patterns for wooden models.

FIG. 55.—A circus parade.

A good circus or "zoo" may be worked out in a variety of materials. Paper, cardboard, clay, and wood all serve well.

To get the best value from the problem it should be as free as possible from copy work. The children should consult the best sources of information at their disposal, which may range all the way from ordinary picture books to natural history and encyclopedia descriptions. They should find out, unaided, as much as possible concerning the animal in question: his size, color, food, home, values, etc.,—the teacher supplementing with interesting and necessary items not at the disposal of the class.

Free-hand cuttings and pencil sketches should be compared with the best pictures obtainable and the real animal whenever possible. Such patterns as are needed should be made by the children themselves. Ready-made patterns will produce better proportioned animals, but more dependent, less observant children also.

METHODS IN DETAIL

Realistic Animals in Three-ply Wood.—Secure necessary items of measurement and decide upon scale. One inch for each foot is best for younger children.

FIG. 56.—Three-ply wooden animals.

Draw rectangle proportioned to the extreme length and height of the animal. Draw into the rectangle a *profile* sketch of the animal, being careful that it comes to the line on each side. *All four feet must* touch the base line. Considerable practice may be needed before a good sketch can be drawn. The animal may be represented as standing, walking, or running, but must be drawn in profile.

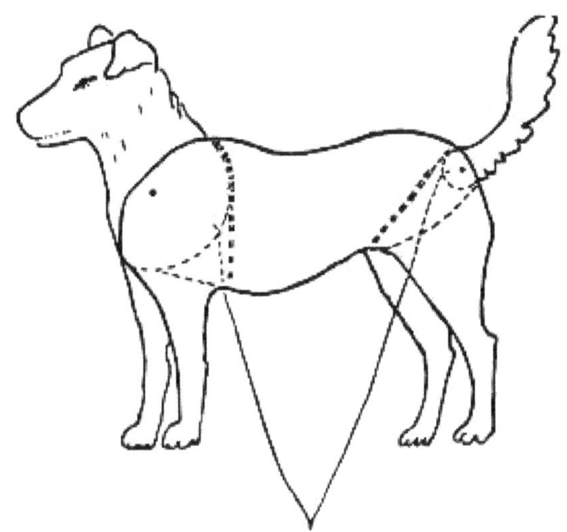

FIG. 57.—Detail of three-ply animals with movable parts.

Cut out the sketch and make by it three patterns: one of the head, body, and tail; one of the body and right legs; one of the body and left legs. Care must be taken to get good lines at shoulder and rump. (See Fig. 56.)

Lay the pattern on the wood so that the grain runs lengthwise of the legs and other frail parts and draw outline carefully. Use basswood one fourth inch thick, or other soft wood.

Saw out the parts with a coping saw. Be careful in sawing to keep the blade in a vertical position in order that the edges may be true.

Nail or glue the parts together. If the animal does not stand perfectly, rub the feet on a piece of sandpaper. Use water color or crayon to give proper color.

Three-ply Animals with Movable Parts.—To make the head movable, saw the part from the body on a curved line, as shown in Fig. 57. Fasten with a single nail through the shoulder. The curved line must be a part of a circle and the nail must be at the center. The edges should be smooth to allow easy action. The tail may be adjusted by a similar plan. The parts may be made to move automatically by suspending a weight on cords which are attached to the movable parts, as shown in Fig. 57. If the weight is to be used, cut off the body part on the double dotted line to allow room for the cords to swing.

A figure of this sort must be fastened on a pedestal or platform which will extend over the edge of the table. A slot must be cut in the pedestal wide enough to allow the cords to swing freely. (See Fig. 56.) The pedestal may be a long board or piece of heavy cardboard which can be tacked to the table or held firm by a clamp, or it may be a thin board fastened to a U-shaped block which is held firm on the edge of the table by a wedge.

Cardboard and Paper Animals that Stand.—For younger children who cannot handle the saw easily cardboard or stiff paper may be used.

FIG. 58.—Notched rest for animals.

To make the animal stand the feet may be tacked to a small piece of wood about one inch square on the end and as long as needed, or a cardboard brace, such as is used on easels, may be glued to the back. A realistic effect is given if the animal is cut with two legs and the brace made to represent the other two, or a piece of cardboard cut as in Fig. 58 may be used as a brace, the body of the animal fitting into the notch.

Clay makes an excellent medium, but it requires more skill in clay than in wood to get an equally good effect. Clay animals should be modeled with a pedestal, and the separations between the two forelegs and the two hind legs merely indicated. If each leg is modeled separately, the figure is likely to be frail.

FIG. 59.—Balancing figures.

Balancing Figures.—Design such figures as a prancing horse or dancing bear and saw from a single piece of wood. A little below the center of the figure insert a curved wire, on the other end of which is a ball of clay or other weight. The wire must be fastened firmly so that it cannot turn. Adjust so that the figure balances.

Figures of people in foreign costumes, children running and jumping, as well as all sorts of animals, are very fascinating problems of this sort. (See Fig. 59.)

Seesaw Figures.—Such groups as two boys chopping wood, two chickens drinking, two dogs tugging at a string, wrestling boys, and similar groups are interesting problems of the seesaw type. (See Fig. 60.)

Detail.—Cut the figures from cardboard. Make with a long pedestal. Color with crayon or water color. Use two light sticks for the seesaw, to which tack one figure in a vertical position and the other on a slant. Fasten to each stick with one tack. If a central figure is used, tack firmly to lower stick. Work the figure by moving the upper stick while the lower one is held firm.

FIG. 60.—Some simple toys.

Toys.—A box of carpenter's scraps of soft wood will supply material for a variety of toys which may be made by the children themselves, thereby more than doubling the fun. A few suggestions are given in detail. The making of these will suggest others. (See Fig. 60.)

Doll's Swing.—A heavy block for a base, two tall uprights, and a crosspiece will make the frame. Make a seat from cardboard or use the end of a small box and suspend from crossbar.

Doll's Teeter.—Use a heavy block for a base. Two uprights with double-pointed tacks or notches in the top. Drive two double-pointed tacks in lower side of teeter board at center. Slip a small rod through the tacks and rest in the notches on the uprights. Suspend a weight by cords from the lower side of the board, adjust until the board balances. The ends of the board should be provided with box seats for the doll's comfort.

Railroad Train.—For cars, saw pieces from a square stick. For engine, use pieces of broomstick or other cylinder. Soft wood is better if obtainable. For wheels, use pieces of small broomstick or dowel rod. (See Fig. 56.)

Let the children study real trains and make the best imitation they can work out.

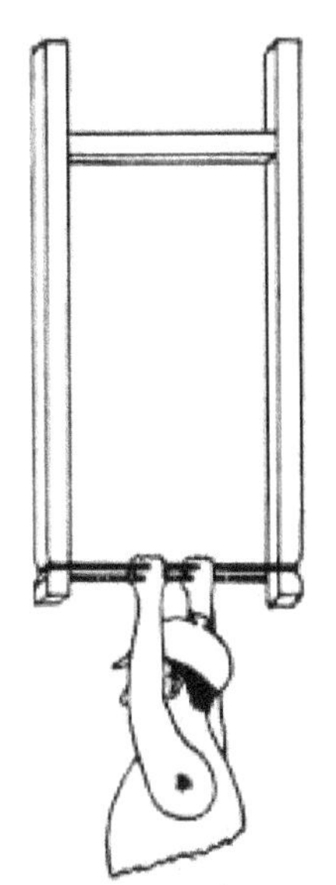

FIG. 61.—Adjusting jumping jack in frame.

Jumping Jacks.—Cut the figure from light weight cardboard. Make head and body in one piece. Cut two arms long enough to reach well above the head. Make the hands very large. Cut two legs either with or without a joint at the knee. Color with crayon or water color.

Fasten the legs and arms to the body with a string tied loosely to allow free movements. Make a frame of two light stiff sticks and a crosspiece fastened between them near the lower end of the sticks. Fasten with a single nail at either end of the crosspiece. Cut notches near the upper ends of the sticks. Fasten the figure to the frame by a stout thread. Use a coarse needle and carry the thread through the hands twice, leaving a loop on each side to slip over the ends of the sticks into the notches. A small block or folded bit of cardboard between the hands to keep them apart will improve the movement of the toy. Adjust the figure so that the threads are parallel when the figure hangs below the inverted frame. (See Fig. 61.) When the frame is held upright, the figure will hang between the sticks and the threads will be crossed. Press the lower ends of the frame together to make the jumping jack perform.

Merry-go-round.—Use a heavy block for a base. Bore a hole in the center and insert a square stick, about 10 in. long. For arms, use two pieces about 3/8 in. thick and 10 in. long. Fasten these together in the form of a cross and nail to the top of the upright with a single nail. An awl may be used to make the hole a little larger than the nail so that the arms will revolve easily.

Suspend a box seat of wood or cardboard from each arm to complete the toy. (See Fig. 59.)

Games.—*Ring Toss.*—Use two square pieces of board at least ½ in. thick, one piece larger than the other. Bore a hole in the center of the smaller piece with a ½-in. auger bit.

For the upright use a stick ½ in. square and about 12 in. long. Whittle the corners of the stick until it fits firmly into the hole in the small board. Nail the small board to the large one.

For the rings use reeds, venetian iron, or hoops from small buckets or cart wheels. Wrap the rings with raffia or yarn. Make at least three rings of varying sizes. (See Fig. 60.)

Playing ring toss and keeping tally makes an excellent number game.

Ten Pins.—From bogus or other heavy paper roll and paste cylinders about three inches in diameter and about twelve inches long. These may be set on end, and any of the common ten pin games played with the help of a soft rubber ball. Keeping tally gives excellent practice in number.

Bean Bag Game.—Draw three circles of different sizes on a large sheet of heavy cardboard. Carefully cut out the circles with a sharp-pointed knife. Mount a picture of some animal on each piece cut out.

Fasten the pieces back in place by a single cloth hinge pasted on the back, and at the lowest part of the circle.

Tack the sheet of cardboard to a light wooden frame to keep it from bending.

Let the frame rest against the wall at a slight angle. Bean bags thrown at the animals will knock them down as they go through the holes. The bean bags should be made by the children. Various number games may be played with bean bags.

CHAPTER IX

HOLIDAYS

The various holidays which come during the year mean so much to little children that they should receive special notice and should suggest the form of handwork to be done at the time.

Thanksgiving suggests attention to harvest products, to be modeled in clay, cut from paper, or drawn with crayon; the making of sand-table scenes showing early New England life in various phases; the making of various utensils and commodities of the primitive home which differ from our own; as, for example, the making of candles, the hour glass, and the sundial.

Christmas suggests the making of toys and all sorts of things suitable for gifts. If the work centers around the Christmas tree, it offers opportunity for coöperation in making trimming such as paper chains, pop-corn strings, etc., as well as individual gifts. If a tree is not obtainable, a box may be dressed up in imitation of Santa's sleigh drawn by cardboard reindeer. Whatever else is done in honor of the visit of St. Nicholas, the spirit of giving should be cultivated by making gifts to some younger or less fortunate groups. Picture books may be made for sick children, doll furniture and other toys for the orphans' home or some family of unfortunates. A sack might arrive a week or two before Christmas accompanied by a telegram from Santa requesting contributions to help him out in some specific way and stating that it would be called for at a certain time. When a "real Santa" calls for the sack, he may leave in its place another containing some unexpected treat for the children themselves. The gifts which the children contribute should be of their own making, that they may have a full sense of real giving and not merely the pleasure of delivering the parcels mother has provided.

Valentine's Day offers an opportunity for developing appreciation of a higher form of art than the shop windows frequently offer, and also investing with pure, sweet sentiment a day which means, in some quarters, only vulgar sentimentality and coarse jests.

Easter offers a similar opportunity for emphasis on the fine things in color and subjects for greeting cards. The season also suggests emphasis on study of budding plants and young animal life by means of cutting, painting, and modeling.

Hero days suggest a variety of forms of handwork, such as picture making with crayons or cuttings, or pictures in three dimensions on the sand table, for intensifying important phases of the hero's life; illustrated stories in booklet form; and the making of "properties" for dramatic representations. These things offer a welcome change from the stereotyped "Speaking day," and stimulate originality and self-reliance.

So much has been written and so many suggestions are constantly being offered in school journals that specific suggestions for *things to make* seem superfluous here.

Individual Problems.—While community problems must form a large part of the handwork in the lower grades, it is desirable to have, from time to time, projects which seek a definite result from each pupil. In the community problem it is possible for the strong pupil to monopolize the values of the work by imposing his ideas upon his fellows and by doing all the work while the slower pupils are getting ready to begin. In the same way it is possible for the lazy pupil to shirk much of his responsibility through the eagerness of his companions. It is therefore necessary to maintain a balance by the use of individual problems of a more definite type. These may often be specific parts of the community problem, but this will not meet all the needs of the case. The special days offer excellent occasion for work of this sort in addition to the coöperative problems which are undertaken by the class as a whole.

CHAPTER X

GENERAL SUGGESTIONS AND SUMMARY

Modification of Outlines.—All the projects outlined in the foregoing pages are capable of modification and adaptation to the needs of several grades. For this reason, in nearly every problem, many more suggestions are offered than will often be applicable in any one instance of its development. The directions are, for the most part, given from the standpoint of the first grade, on the principle that it is easier to add to the detail of a problem than to simplify it. On the other hand, the directions are not generally specific in detail, in order to prevent as far as possible a mechanical copying of any project.

Emphasis on Self-expression.—It is desired to place special emphasis upon the point that each project undertaken, if it is to reach its highest value, must come as fully as possible from the children themselves and be to the very fullest extent *their self-expression.*

Not any house described in this book, nor any house seen in another schoolroom, nor even the house which I, as teacher, plan in detail, will be most valuable to my class; but rather *our house*, which *we, teacher and pupils* working together, evolve to suit our own needs and fancies, using suggestions gathered from every available source, but adapting them to our own needs.

Self-directed Activity and Discipline.—The terms "self-directed activity" and "self-expression" must not be confounded with the idea of letting the children do as they please in any random and purposeless fashion. If one were to start out to escort a group of children to a certain hilltop, it is quite probable that some of them would run part of the way. Others would walk in twos and threes, and these would change about. They would halt to look at things that attracted their attention. The leader would halt them to observe some interesting point which they might otherwise miss. Should any of them wander from the right path the leader would call them back, and any frail child would be helped over the hard places. Yet with all this freedom the group might move steadily forward and reach the hilltop in due time.

All progress up the hill of knowledge should follow a similar plan. The teacher should have a very definite idea of the end to be attained. The children should work with a purpose, and that purpose should be of such immediate interest to them that they would be anxious to attain it. They would then work earnestly, and discipline would settle itself. Handwork projects should be sufficiently simple to allow each worker to see his way through, or at least find his way without waiting for directions at each step. Instead of a blind following of such directions the worker should at all times feel himself the master of his tools and materials and be able to make them obey his impulse and express his idea. This attitude toward work can be secured only when the work is kept quite down to the level of the child's ability and appreciation. Only by this means can we hope to establish the inspiring and strengthening "habit of success."

Introduction of New Methods.—The question arises, How shall work of this sort be adapted to a course of study which is already full and does not provide time for handwork? Handwork takes more time than bookwork, and children evolve plans but slowly. If the teacher waits for the children to evolve plans and then carry them out on their own responsibility, the quantity of work produced will be small and the quality poor compared with the results gained by other methods.

The freer method must be justified, not by its tangible results, but by its value as a means of individual development. If it is true that

"One good idea known to be
thine own
Is worth a thousand gleaned
from fields by others sown,"

then it follows that a small quantity of crude work may often represent greater genuine growth than a larger quantity of nicely finished work, if the latter has been accomplished by such careful dictation that individual thought on the part of the pupils was unnecessary.

Common sense is the best guide in introducing a new method of work. Any sudden transition is likely to be disastrous. Responsibility in new fields should be shifted from teacher to pupils as rapidly as they are able to carry it, but it should never be transferred in wholesale fashion. This is especially true of a class that is accustomed to wait for the teacher's permission or command in all the small details of schoolroom life, such as speaking, moving about the room, etc.

The freer methods may be introduced by either of two plans. In carrying through the first sand-table project, for example, the teacher may plan the details quite as definitely as is her custom in general work, assign each part to a particular pupil, and guide his execution of it as far as necessary. With each succeeding project more and more freedom may be granted, as the children become accustomed to community work and learn how to use the materials involved. Or, the work may be introduced by allowing two or three very trustworthy pupils to work out, quite alone, some simple project which will appeal to the entire class as very desirable. Other projects may be worked out by other pupils as they show themselves worthy of trust. Such a plan sets a premium upon independence and ability to direct one's own actions, and has a beneficial effect upon general discipline. Each individual teacher must follow the plan which best accords with her individual habits and the conditions under which she works. No rule can be rated as best under any and all circumstances.

New and Different Projects.—Teachers frequently spend time and nerve force seeking new projects supposedly to stimulate the interest of the children. Often a careful examination into the true motives back of the search would prove that it is not so much to stimulate the interest of the children as to call forth the admiration of other teachers. Because a house was built last year does not hinder the building of another this year. If the children are allowed ample freedom, the houses will not be alike. If the teacher is centering her interest in the development of the children and not in the things the children make, the projects will always be new because worked out in a new way by a different group of children. Monotony comes about through the teacher's attempt to plan out details and impose them upon the children, a process quite similar to the use of predigested foods.

Quality of Work.—Methods such as outlined above are sometimes criticized because of the crudity of the results. It is sometimes argued that the crude work establishes low standards and that better finished work of a more useful type is more desirable in school projects. Certainly everything which is done in school should be useful. School years are too precious to be wasted, in any degree, on a thing which is useless. But it is important to have a right standard for measuring the usefulness of a project. Since it is the child's interest and effort which are to be stimulated, his work must be useful from his point of view. The things that he works upon must be valuable to him personally. It is not enough for the teacher to be satisfied with the value of the subject matter. It must, as far as possible, be self-evident to the child himself.

In the growing period a child is always anxious to excel himself and attain a higher level, nearer the adult standards. He measures his growth, not only in inches, but in ability to run faster, jump farther, count higher, and so on. So long as he is stimulated by an interesting motive he puts forth his best effort. It is only when we set him tasks and demand blind obedience that he lags. If his crude work represents his best effort, honestly put forth, he will, and he does, desire to do something better each time he tries. If he is permitted to work freely upon projects of immediate interest to him, he not only becomes familiar with various materials and the purposes they may serve, but he also begins to realize his inability to make them always obey his impulse. As soon as he discovers that there are better and easier ways of working which bring about more satisfactory results, he is anxious to learn the tricks of the trade; and he comes to the later, more technical courses in handwork, not only with more intelligence, but also with an appreciation of their value which is reflected in the quality of his work.

Summary.—The last word, as the first in this little book, would stress the fact that it is always possible to improve present conditions.

Activity is an essential factor in a child's development in school as well as out. Handwork is an important phase of this necessary activity. Neither lack of time, scarcity of material, nor lack of training on the part of the teacher is a sufficient excuse for failure to use some handwork in every school. Much can be accomplished with materials which are to be found anywhere, without using more time than is ordinarily devoted to the subject, and with better results, if we will but realize that educative handwork is not confined to the making of a few books, boxes, mats, or baskets after a prescribed pattern, however good in themselves these may be, but is also a means through which we may teach other subject matter.

We not only learn to do by doing, but we come to *know* through trying to *do.* And we often learn more through our failures than through our successes. We defraud the children if we deprive them of this important factor in their development. Any teacher who is willing to begin with what she has and *let the children do* the best they can with it, will find unexpected resources and greater opportunities at every hand.

Let us not allow ourselves to grow disheartened through vain wishes for the impossible or for the advantages of some other field, but attack our own with vigor and determination; for

> "The common problem, yours,
> mine, every one's
> Is—not to fancy what were fair
> in life
> Provided it could be—but,
> finding first
> What may be, then find how to
> make it fair
> Up to our means."

REFERENCES

DEWEY—The School and the Child; School and Society; The Child and the Curriculum.

O'SHEA—Dynamic Factors in Education.

SCOTT—Social Education.

DOPP—The Place of Industries in Elementary Education.

BONE—The Service of the Hand in the School.

SARGENT—Fine and Industrial Arts.

ROW—The Educational Meaning of Manual Arts and Industries.

CHARTERS—Methods of Teaching.

BAGLEY—The Educative Process.

RUSSELL—The School and Industrial Life. Educational Review, Dec. 1909.

SYKES AND BONSER—Industrial Education. Teachers College Record, Sept. 1911.

BENNETT—The Place of Manual Arts in Education. Educational Review, Oct. 1911.

RICHARDS—Handwork in the Primary School. Manual Training Magazine, Oct. 1901.

REFERENCES FOR CLASSROOM USE

Coping Saw Work — JOHNSTON

School Drawing — DANIELS

Little Folks Handy Book — BEARD

World at Work Series — DUTTON

Big People and Little People of Other Lands — SHAW

How We Are Fed — CHAMBERLAIN

How We Are Clothed — CHAMBERLAIN

How We Are Sheltered	CHAMBERLAIN
Continents and their People	CHAMBERLAIN
How the World is Fed	CARPENTER
How the World is Clothed	CARPENTER
How the World is Housed	CARPENTER
Around the World Series	TOLMAN
Youth's Companion Series	LANE
The Bird Woman	CHANDLER
The Tree Dwellers	DOPP
The Early Cave Men	DOPP
The Later Cave Men	DOPP

Footnotes:

[1] In scoring cardboard cut about halfway through the board on the *outside* of the fold.

[2] See Scott's "Social Education."

[3] See Riverside Primer.

Milton Keynes UK
Ingram Content Group UK Ltd.
UKHW050650260624
444769UK00004B/200

9 789362 098856